Instructor's
Resource Manual

DISCRETE
MATHEMATICS

Instructor's
Resource Manual

third edition
DISCRETE
MATHEMATICS

Kenneth A. Ross
Charles R.B. Wright
Department of Mathematics
University of Oregon

 PRENTICE HALL, Englewood Cliffs, New Jersey 07632

Printed in the United States of America
10 9 8 7 6 5 4 3

ISBN   0-13-218165-7

Prentice-Hall International (UK) Limited, *London*
Prentice-Hall of Australia Pty. Limited, *Sydney*
Prentice-Hall Canada Inc., *Toronto*
Prentice-Hall Hispanoamericana, S. A., *Mexico*
Prentice-Hall of India Private Limited, *New Delhi*
Prentice-Hall of Japan, Inc., *Tokyo*
Simon & Schuster Asia Pte. Ltd., *Singapore*
Editora Prentice-Hall do Brasil, Ltda., *Rio de Janeiro*

CONTENTS

PREFACE   vii

Chapter 1   1

Chapter 2   16

Chapter 3   33

Chapter 4   50

Chapter 5   70

Chapter 6   83

Chapter 7   105

Chapter 8   125

Chapter 9   143

Chapter 10   157

Chapter 11   170

Chapter 12   185

Chapter 13   212

PREFACE

This manual contains answers for all of the exercises in the text that call for explicit responses. In a number of instances we have elaborated on the answers that appear in the back of the text. In addition, we have preceded the solutions for each chapter by some suggestions and comments that we hope will be helpful to the instructor.

We have tried to divide the material in the text so that each section corresponds to a single class period. Experience has shown, however, that some sections really deserve two days, either because they contain a lot of new ideas or because students will need more than the usual amount of expert guidance. We have indicated in our chapter suggestions those sections that we think could use extra time. For quick reference, they are: §§ 4.2 [induction], 7.3 [depth-first search], 9.3 [mean and variance], 11.1 [partial orders] if you skip Chapter 10, 11.2 [special linear orders], and 12.8 [rings and fields]. Section 1.6 [big-oh notation] might also warrant an extra day.

Instructor's
Resource Manual

DISCRETE
MATHEMATICS

CHAPTER 1

This chapter goes over elementary properties of sets, and establishes notation for sets and sequences. Much of the material will be old news for the students, and it may be hard to convince them that they don't know it all already. One of the aims of this chapter is to set the tone of the book by emphasizing the absolute necessity of careful definitions of even the most common terms.

In § 1.1 students have trouble with $\mathscr{P}(S)$ and Σ^*, both of which will be important later on. Now is the time to look at lots of examples, to get comfortable with the notation and to see how natural the ideas are.

Section 1.2 contains a number of proofs, and many of its exercises say "prove," yet we don't discuss proofs formally until Chapter 2. For now, students should think of "prove" in the sense of the message to the student in the book's preface. Teachers, on the other hand, should be going over proofs, such as those in Examples 3-5, very carefully in class, as models. In the exercises, "determine" means list or describe the set unambiguously. If you assign Exercise 15, you may want to explain the term "counterexample."

Section 1.3 is pretty standard. We tend to think of functions as rules, but from time to time also want to consider them as sets of ordered pairs, especially later on in Chapter 3 when we look at relations. Many of the examples seem more like the functions encountered in calculus than those associated with discrete mathematics. We've made these choices not to stress calculus and real-valued functions but to try to connect abstract ideas, such as one-to-one, with familiar geometric settings. It also helps to have students think of functions in terms of hand-held calculators. If the square-root function were two-valued, what would the calculator display do? And how about inverse trigonometric functions?

In § 1.4 it is not necessary to prove the theorem in class; just give plenty of examples. Experience tells us that students have trouble with both the notation and the concept of preimage. The notion of inverse also seems slippery for some. To keep the two concepts distinct, we have introduced the notation $f^\leftarrow(B)$ for non-invertible functions f. Examples help. Examples 6 and 7 are important for later work on equivalence relations in Chapter 3. The last paragraph in this section is deceptively innocent. We do want students to think of this powerful observation as completely natural. It will get fairly heavy use later on.

The basic ideas of sequences in § 1.5 seem not to cause trouble. We've used this section to work in sigma notation and introduce the idea of growth rate, using some popular sequences for illustration.

Section 1.6 contains a careful account of big-oh notation, which we use extensively later on to estimate algorithm complexity. Students may already have encountered this notation in their computer science classes, where the treatment is likely to have been off-hand and confusing. They may also have learned some tricks using limits to calculate estimates for sequences of the form $f(n)/g(n)$. Because we naturally want the smallest m such that $f(n) = O(n^m)$, students may think that $f(n) = O(n^m)$ means that $n^m = O(f(n))$ too. Try to watch for this misconception.

Emphasize that the big-oh notation is very handy, but that it is not a license to be sloppy. Quite the contrary, it gives a precise way of describing rough estimates. Exercises 11 and 12 illustrate the second way we use the notation. Exercise 17 connects $\log n$ with the number of digits in n, an important observation in computer science.

Section 1.1

1. (a) 0, 5, 10, 15, 20, say.
   (b) 3, 5, 7, 9, 11, say.
   (c) Ø, {1}, {2, 3}, {3, 4}, {5}, say.
   (d) 1, 2, 4, 8, 16, say.
   (e) 1, 1/2, 1/3, 1/4, 1/73, say.
   (f) 1/2, 1/3, 1/4, 1/5, 17/73, say.
   (g) 1, 2, 4, 16, 18, say.

2. (a) 1, 1/2, 1/3, 1/4.
   (b) 0, 2, 6, 12.
   (c) 1/4, 1/16, 1/36, 1/64, 1/100.
   (d) 1, 3.

3. (a) λ, a, ab, cab, ba, say.
   (b) {λ, a, b, aa, ab, ba, bb} is the complete set.
   (c) aaaa, aaab, aabb, etc.
   The sets in parts (a) and (b) contain the empty word λ.

4. (a) {3}.　(b) {-3, 3}.　(c) {-3, 3}.　(d) {4, 5, 6}.
   (e) {-6, -5, -4, 4, 5, 6}.　(f) Ø.　(g) Ø.　(h) Ø.　(i) Ø.
   (j) {1, 4, 7, 10, 13, 16, 19}.　(k) {2, 3, 5, 7, 11, 13}.

5. (a) 0.　(b) 74.　(c) 138.　(d) 67.　(e) 73.
   (f) ∞.　(g) 0.　(h) 2.　(i) ∞.　(j) $2^4 = 16$.
   (k) ∞.　(l) ∞.　(m) ∞.　(n) 1.　(o) ∞.

6. (a) 2. (b) ∞. (c) ∞. (d) 3. (e) ∞.
    (f) $1 + 3 + 9 + 27 + 81 = 121$.

7. $A \subseteq A$, $B \subseteq B$, C is a subset of A, and C, D are subsets of A, B and D.

8. (a) False. (b) True. (c) True. (d) True. (e) False.
    (f) False. (g) False. (h) True. (i) True.

9. (a) aba is in all three and has length 3 in each.
    (b) bAb is in Σ_3^* and has length 2.
    (c) cba is in Σ_1^* and length(cba) = 3.
    (d) cab has length 3 in Σ_1^* and length 2 in Σ_2^*.
    (e) caab is in Σ_1^* with length 4 and is in Σ_2^* with length 3.
    (f) baAb has length 3 in Σ_3^*.

10. In the first case there would be infinitely many words before ba: all the words beginning with a and then b. In the second case there would be $1 + 2 + 4 + 8 + 16 = 31$ words beginning with a, and then b before ba.

11. (a) Yes. (b) No: consider $\Sigma = \{a, b, Ab\}$ and abbAb.
    (c) Delete first letters from the string until no longer possible. If λ is reached, the original string is in Σ^*. Otherwise, it isn't.

Section 1.2

1. (a) $\{1, 2, 3, 5, 7, 9, 11\}$. (b) $\{3\}$. (c) $\{1, 5, 7, 9, 11\}$. (d) $\{1, 9\}$.
    (e) $\{3, 6, 12\}$. (f) $\{3, 4, 5, 7, 8, 11\}$. (g) 16.

2. (a) $\{2\}$, $\varnothing$, $\mathbb{P}$, $\mathbb{P}$.
    (b) $\varnothing$, $\{1\}$, $\{2\}$, $\{3\}$, $\{1, 2\}$, $\{1, 3\}$, $\{2, 3\}$, $\{1, 2, 3\}$.
    (c) $A \oplus B$, $A \oplus C$, $C \setminus A$.

3. (a) $[2,3]$. (b) $[0,6]$. (c) $[0,2)$. (d) $[0,2) \cup (3,6]$.
    (e) $(-\infty,0) \cup (3,\infty)$. (f) $\varnothing$.

4. (a) $\{a, b, aa, bb\}$, $\{aaa, bbb\}$, $\{\lambda, ab, ba\}$, $\{\lambda, ab, ba, aaa, bbb\}$.
    (b) $\{aa, bb, aaa, bbb\}$, $\{aa, ab, ba, bb\}$, Σ^*, the set of words of length 2 or more, except for aa, bb, aaa and bbb.

3

(c) $\{\lambda, a, b\}$, $\{a, b\}$, $\emptyset$.　(d) $\emptyset$, $\{a\}$, $\{b\}$, $\{a, b\}$.　(e) 4.

5. (a) $\emptyset$.　　　　(b) All words whose length is not 2.
   (c) $\emptyset$.　　　　(d) Same as part (b).
   (e) $\{\lambda, ab, ba\}$.　(f) $\{\lambda\}$.
   (g) $B^c \cap C^c$ and $(B \cup C)^c$ are equal by a DeMorgan law [or by calculation] as are $(B \cap C)^c$ and $B^c \cup C^c$.

6. (a) False. For example, let $A = \emptyset$, $B = C = \{a\}$.
   (b) True.　(c) True.　(d) True.　(e) False. Use any A and B.

7. $A \oplus A = \emptyset$ and $A \oplus \emptyset = A$.

8. (a)

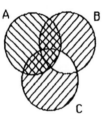
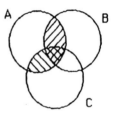

$A \cap (B \oplus C)$ is double-hatched

$(A \cap B) \oplus (A \cap C)$ is single-hatched

(b)

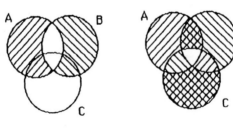

$A \oplus B$ is hatched

$(A \oplus C) \cup (B \oplus C)$ is hatched

9. $(A \cap B \cap C)^c = (A \cap B)^c \cup C^c$ and $(A \cap B)^c = A^c \cup B^c$ by DeMorgan law 9b. Now substitute.

10. (a) Every x in $A \cap B$ is in both A and B, so in particular is in A. Hence $A \cap B \subseteq A$.

Every x in A is in A or B so is in $A \cup B$. Thus $A \subseteq A \cup B$.

[Students may be puzzled about what to write for such an "obvious" question.]

(b) Consider an element x of A. Since $A \subseteq B$, $x \in B$. Since $A \subseteq C$, $x \in C$. Thus x is in both B and C, i.e., $x \in B \cap C$. Hence $A \subseteq B \cap C$.

(c) Consider an element x of $A \cup B$. Then x is in A or x is in B. In either case, $x \in C$ since $A \subseteq C$ and $B \subseteq C$ by hypothesis. Thus we have $A \cup B \subseteq C$.

(d) If $A \subseteq B$ and if $x \in B^c$, then x is not in B so x is surely not in A, i.e., $x \in A^c$. Thus if $A \subseteq B$ then $B^c \subseteq A^c$.

Conversely, if $B^c \subseteq A^c$ and if $x \in A$, then x is not in A^c so x is not in B^c and thus x must be in B. Thus if $B^c \subseteq A^c$ then $A \subseteq B$.

Alternatively, if $B^c \subseteq A^c$ then by what was just shown and law 6 we have $A = (A^c)^c \subseteq (B^c)^c = B$.

11. (a) (a,a), (a,b), (a,c), (b,a), etc. There are nine altogether.

(b) (a,a), (a,b), (a,d), (b,a), (b,b), (b,d), (c,a), (c,b), (c,d).

(c) (a,a), (b,b).

12. (a) 15, 15.

(b) (0,2), (0,4), (1,2), (1,4), (2,4), (3,4).

(c) (0,1), (0,2), (0,3), (0,4), (2,3), (2,4).

(d)
```
4  •  •  •  •  •
2  ○  •  •  •  •
0  ○  ○  ○  •  •
   0  1  2  3  4

      S × T
```

(e)
```
4  ○  •  •
3  ○  •  •
2  ○  •  •
1  ○  ○  •
0  ○  ○  ○
   0  2  4

      T × S
```

(f) There are none.

13. (a) (0,0), (1,1), (2,2), ... , (6,6), say.

(b) The set is infinite. (0,2), (6,5), (2,3) are examples.

(c) (6,1), (6,2), (6,3), ... , (6,7), say.

(d) The set is infinite. (3,5), (73,3), (3,3) are examples.

(e) (1,3), (2,3), (3,3), (3,2), (3,1).

(f) The set is infinite. (1,1), (0,0), (10,100) are examples.

14.

See *Mathematics Magazine 58* (Sept. 1985), p. 251 for five rectangles forming a Venn diagram.

15. (a) False. Try $A = \emptyset$.

(b) False. Try $A = B \neq \emptyset$ and $C = \emptyset$ or try $A =$ the universe U.

(c) True. Show $x \in B$ implies $x \in C$ by considering two cases: $x \in A$ and $x \notin A$. Similarly, $x \in C$ implies $x \in B$.

(d) True. If $A \cup B \subseteq A \cap B$, then using Exercise 10(a) we have $A \subseteq A \cup B \subseteq A \cap B \subseteq B$, so $A \subseteq B$. Similarly $B \subseteq A$.

(e) The hint to part (c) also applies here.

16. (a) Suppose that $A \subseteq B$. Consider $x \in A \cup B$. Then $x \in A$ or $x \in B$. Since $A \subseteq B$, $x \in B$ in either case. Thus $A \cup B \subseteq B$. Since $B \subseteq A \cup B$ always holds [by Exercise 10(a)], $A \cup B = B$ if $A \subseteq B$.

Next suppose that $A \cup B = B$ and consider $x \in A$. Then $x \in A \cup B$, so $x \in B$ by assumption. Thus $A \subseteq B$ if $A \cup B = B$.

(b) The outline of the proof is similar to that of part (a). Suppose that $A \subseteq B$. Consider $x \in A$. Since $A \subseteq B$, we also have $x \in B$, and so $x \in A \cap B$. Thus $A \subseteq A \cap B$. Since $A \cap B \subseteq A$ always holds [Exercise 10(a) again], $A \cap B = A$ if $A \subseteq B$.

Next suppose that $A \cap B = A$ and consider $x \in A$. Then $x \in A \cap B$ and so $x \in B$. Thus $A \subseteq B$ if $A \cap B = A$.

17. (a) Any example with $A \neq B$ shows the failure.

(b) Any example with $A \cap C \neq \emptyset$ shows the failure.   (c) The Venn diagrams are

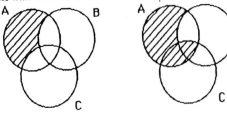

$(A \setminus B) \setminus C$ $A \setminus (B \setminus C)$

Section 1.3

1. (a) $f(3) = 27$, $f(1/3) = 1/3$, $f(-1/3) = 1/27$, $f(-3) = 27$.

   (b)

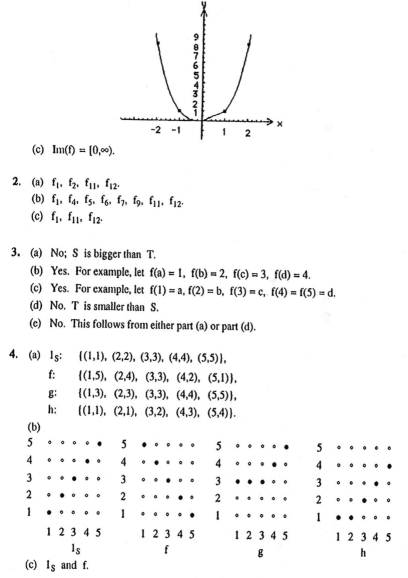

   (c) $Im(f) = [0, \infty)$.

2. (a) f_1, f_2, f_{11}, f_{12}.
   (b) f_1, f_4, f_5, f_6, f_7, f_9, f_{11}, f_{12}.
   (c) f_1, f_{11}, f_{12}.

3. (a) No; S is bigger than T.
   (b) Yes. For example, let $f(a) = 1$, $f(b) = 2$, $f(c) = 3$, $f(d) = 4$.
   (c) Yes. For example, let $f(1) = a$, $f(2) = b$, $f(3) = c$, $f(4) = f(5) = d$.
   (d) No. T is smaller than S.
   (e) No. This follows from either part (a) or part (d).

4. (a) 1_S: $\{(1,1), (2,2), (3,3), (4,4), (5,5)\}$,
      f: $\{(1,5), (2,4), (3,3), (4,2), (5,1)\}$,
      g: $\{(1,3), (2,3), (3,3), (4,4), (5,5)\}$,
      h: $\{(1,1), (2,1), (3,2), (4,3), (5,4)\}$.

   (b)

   (c) 1_S and f.

7

5. (a) $f(2,1) = 2^2 3^1 = 12$, $f(1,2) = 2^1 3^2 = 18$, etc.

   (b) If not, $2^m 3^n = 2^{m'} 3^{n'}$ for some $(m,n) \neq (m',n')$. Then $m \neq m'$ [why?]. Say $m < m'$. Divide both sides by 2^m to get a number that is both odd and even, a contradiction.

   (c) Consider 5, for instance. 5 is not in Im(f).

   (d) For example, $g(2,0) = g(0,1)$. In general, $g(m,n) = g(m',n')$ if and only if $m + 2n = m' + 2n'$.

6. (a) 1_N, f, g.　　　　　(b) 1_N, g, k.

7. (a) Pick b_0 in B. For every $a \in A$, $PROJ(a,b_0) = a$ and so every a in A is in the image of PROJ. That is, PROJ maps $A \times B$ onto A.

   (b) If B has just one element, PROJ is one-to-one. Otherwise, if $b \neq b'$ with $b \in B$ and $b' \in B$, and if $a \in A$, then $PROJ(a,b) = PROJ(a,b')$, so PROJ is not one-to-one.

8. (a) $L(w_1) = 3$, $L(w_2) = 6$ and $L(w_3) = 0$. Note that w_3 is the empty word λ which uses no letters at all and so has length 0.

   (b) No. For example, $cab \neq abc$ but $L(cab) = L(abc)$.

   (c) Yes. $0 = L(\lambda)$, $1 = L(a)$, $2 = L(aa)$, $3 = L(aaa)$, etc. Indeed, $n = L(a^n)$ where a^n stands for the string of n a's.

   (d) There are nine: aa, ab, ac, ba, bb, bc, ca, cb, cc.

9. $\{n \in \mathbb{Z} : n \text{ is even}\}$.

10. Both equal 2.

11. (a) $f \circ g \circ h(x) = (x^8 + 1)^{-3} - 4(x^8 + 1)^{-1}$.

    (b) $f \circ h \circ g(x) = (x^2 + 1)^{-12} - 4(x^2 + 1)^{-4}$.

    (c) $h \circ g \circ f(x) = [(x^3 - 4x)^2 + 1]^{-4}$.

    (d) $f \circ f(x) = (x^3 - 4x)^3 - 4(x^3 - 4x)$.

    (e) $g \circ g(x) = (x^2 + 1)^2 / [1 + (x^2 + 1)^2]$.

    (f) $h \circ g(x) = (x^2 + 1)^{-4}$.

    (g) $g \circ h(x) = (x^8 + 1)^{-1}$.

12. If $(g \circ f)(s) = (g \circ f)(s')$, then $g(f(s)) = g(f(s'))$. Since g is one-to-one, $f(s) = f(s')$. Then since f is one-to-one, $s = s'$. Thus $g \circ f$ is one-to-one.

    Alternatively, suppose $s \neq s'$. Since f is one-to-one, $f(s) \neq f(s')$. Since g is one-to-one, $(g \circ f)(s) = g(f(s)) \neq g(f(s')) = (g \circ f)(s')$.

13. Since $g \circ f: S \to U$ and $h: U \to V$, the composition $h \circ (g \circ f)$ is defined and maps S to V. A similar remark applies to $(h \circ g) \circ f$. Show that the functions' values agree at each $x \in S$.

14. The values of the function $1_{\mathbb{R}}$ can be calculated by doing nothing at all. They can also be considered as produced by keying in entries.

15. (a) $1, 0, -1$ and 0.

(b) $9, 10, 1, 0$.

(c) $g \circ f$ is the characteristic function of $\mathbb{Z} \setminus E$. $f \circ f(n) = n - 2$ for all $n \in \mathbb{Z}$.

(d) $g \circ g(2m) = g(g(2m)) = g(1) = 0 = g(2m - 1) = g \circ f(2m)$, and $g \circ g(2m + 1) = g(0) = 1$ $= g(2m) = g \circ f(2m + 1)$. Thus $g \circ g = g \circ f$. Also $f \circ g(2m) = f(1) = 0 = -0 = -g(2m - 1) =$ $-g \circ f(2m)$, and $f \circ g(2m + 1) = f(0) = -1 = -g(2m) = -g \circ f(2m + 1)$. Hence $f \circ g = -g \circ f$.

Section 1.4

1. (a) $f^{-1}(y) = (y - 3)/2$. (b) $g^{-1}(y) = \sqrt[3]{y + 2}$.

   (c) $h^{-1}(y) = 2 + \sqrt[3]{y}$. (d) $k^{-1}(y) = (y - 7)^3$.

2. (a) $(0, \infty)$, $\mathbb{R}$, $[0, \infty)$, $(-\infty, 0) \cup (0, \infty)$.

   (b) The function $f(x) = 1/x$ is its own inverse. The inverse of the square-root function is the square function with domain $[0, \infty)$, but not with domain $\mathbb{R}$.

   (c) The functions given by x^2, $\sqrt{x}$ and $1/x$ commute with each other on their common domain $(0, \infty)$, but not with $\log x$.

   (d) $\mathbb{R}$, $\mathbb{R}$, $\mathbb{R} \setminus \{(2n + 1)\pi/2 : n \in \mathbb{Z}\}$; none; none.

3. (a) All of them; verify this.

   (b) Each has the same values on both (m, n) and (n, m), for example.

   (c) $\text{SUM}^{\leftarrow}(4)$ has 5 elements, $\text{PROD}^{\leftarrow}(4)$ has 3 elements, $\text{MAX}^{\leftarrow}(4)$ has 9 elements, and $\text{MIN}^{\leftarrow}(4)$ is infinite.

4. (a) Given $A \in \mathscr{P}(\mathbb{N})$, we have $\text{UNION}(A, \emptyset) = \text{INTER}(A, \mathbb{N}) = \text{SYM}(A, \emptyset) = A$.

   (b) Each has the same values on both (A, B) and (B, A), for example.

   (c) $\text{UNION}^{\leftarrow}(\emptyset) = \{(\emptyset, \emptyset)\}$, with 1 member.

   $\text{INTER}^{\leftarrow}(\emptyset)$ consists of all pairs of disjoint sets, an infinite subset of $\mathscr{P}(\mathbb{N}) \times \mathscr{P}(\mathbb{N})$.

   $\text{SYM}^{\leftarrow}(\emptyset) = \{(A, A) : A \in \mathscr{P}(\mathbb{N})\}$, an infinite set.

   $\text{UNION}^{\leftarrow}(\{0\}) = \{(\emptyset, \{0\}), (\{0\}, \emptyset), (\{0\}, \{0\})\}$, with 3 members.

   $\text{INTER}^{\leftarrow}(\{0\})$ and $\text{SYM}^{\leftarrow}(\{0\})$ are infinite sets.

5. (a) $f(0) = 1$, $f(1) = 2$, $f(2) = 3$, $f(3) = 4$, $f(4) = 5$, $f(73) = 74$.

(b) $g(0) = 0$, $g(1) = 0$, $g(2) = 1$, $g(3) = 2$, etc.

(c) For one-to-oneness, observe that if $f(n) = f(n')$, then $n = f(n) - 1 = f(n') - 1 = n'$.
f does not map onto $\mathbb{N}$ because $0 \notin \text{Im}(f)$.

(d) g maps onto $\mathbb{N}$ because $g(n + 1) = n$ for each $n \in \mathbb{N}$. g is not one-to-one
because $g(0) = g(1) = 0$.

(e) $g(f(n)) = \max\{0,(n + 1) - 1\} = n$, but $f(g(0)) = f(0) = 1$.

6. (a) 0, 0, 1, 1, 2, 36.

(b) $g(f(n)) = (2n)/2 = n$, since $2n$ is even. $f(g(n)) = n - 1$ for odd n. A single example
is enough.

7. (a) $(f \circ f)(x) = 1/(1/x) = x$. (b) - (d) are similar verifications.

8. $\chi_A^{\leftarrow}(1) = A$; $\chi_A^{\leftarrow}(0) = S \setminus A$.

9. Since f and g are invertible, the functions $f^{-1}: T \to S$, $g^{-1}: U \to T$ and $f^{-1} \circ g^{-1}: U \to S$
exist. So it suffices to show $(g \circ f) \circ (f^{-1} \circ g^{-1}) = 1_U$ and $(f^{-1} \circ g^{-1}) \circ (g \circ f) = 1_S$.

10. To show that f^{-1} is invertible we must show that there is a function $(f^{-1})^{-1}$ such that
$(f^{-1})^{-1} \circ f^{-1} = 1_T$ and $f^{-1} \circ (f^{-1})^{-1} = 1_S$. The function $(f^{-1})^{-1} = f$ has these two properties, since
$f \circ f^{-1} = 1_T$ and $f^{-1} \circ f = 1_S$. If some function $g: S \to T$ satisfies $g \circ f^{-1} = 1_T$, then $g = g \circ 1_S$
$= g \circ (f^{-1} \circ f) = (g \circ f^{-1}) \circ f = 1_T \circ f = f$. Thus f is the only candidate for $(f^{-1})^{-1}$. [Inverses, when
they exist, are always unique.]

11. (a) Prove: if $s_1, s_2 \in S$ and $f(s_1) = f(s_2)$, then $s_1 = s_2$. The proof will be very short.
Indeed, if $f(s_1) = f(s_2)$, then $s_1 = (g \circ f)(s_1) = g(f(s_1)) = g(f(s_2)) = s_2$.

(b) For each s in S, g maps $f(s)$ onto s.

12. (a) If $f(x,y) = f(x',y')$, then $x + y = x' + y'$ and $x - y = x' - y'$. Hence $2x = (x + y) +$
$(x - y) = 2x'$ and $2y = (x + y) - (x - y) = 2y'$, so that $x = x'$, $y = y'$ and $(x,y) = (x',y')$.

(b) For each $(a,b) \in \mathbb{R} \times \mathbb{R}$, $f(\frac{a+b}{2}, \frac{a-b}{2}) = (a,b)$.

(c) $f^{-1}(a,b) = (\frac{a+b}{2}, \frac{a-b}{2})$.

(d) $f \circ f^{-1} = 1_{\mathbb{R} \times \mathbb{R}}$; $(f \circ f)(x,y) = (2x,2y)$.

13. (a) Suppose $t \in f(f^{\leftarrow}(B))$. Then $t = f(s)$ for some $s \in f^{\leftarrow}(B)$. $s \in f^{\leftarrow}(B)$ means that $f(s) \in B$. So $t \in B$. This works for any t, so $f(f^{\leftarrow}(B)) \subseteq B$.

(b) The statement $B \subseteq f^{\leftarrow}(f(A))$ means $f(b) \in f(A)$ for every b in B, i.e., $f(B) \subseteq f(A)$, so clearly $A \subseteq f^{\leftarrow}(f(A))$.

(c) First, $f(f^{\leftarrow}(B_1) \cap f^{\leftarrow}(B_2)) \subseteq f(f^{\leftarrow}(B_1)) \cap f(f^{\leftarrow}(B_2)) = B_1 \cap B_2$, so
$$f^{\leftarrow}(B_1) \cap f^{\leftarrow}(B_2) \subseteq f^{\leftarrow}(B_1 \cap B_2).$$
In the other direction, $f(f^{\leftarrow}(B_1 \cap B_2)) = B_1 \cap B_2 \subseteq B_1$, so $f^{\leftarrow}(B_1 \cap B_2) \subseteq f^{\leftarrow}(B_1)$, and similarly $f^{\leftarrow}(B_1 \cap B_2) \subseteq f^{\leftarrow}(B_2)$. Thus $f^{\leftarrow}(B_1 \cap B_2) \subseteq f^{\leftarrow}(B_1) \cap f^{\leftarrow}(B_2)$.

(d) Equality holds if and only if $B \subseteq f(S)$.

14. (a) False. For instance, $A_1 \cap A_2$ can be empty while $f(A_1) \cap f(A_2)$ is nonempty. Example: $f: \mathbb{Z} \to \mathbb{Z}$ defined by $f(n) = n^2$, $A_1 = \mathbb{P}$, $A_2 = \{-n : n \in \mathbb{P}\}$.

(b) and (c) are false. See the answer to part (a).

15. (a) The first sentence shows that the f in the example cannot be one-to-one. At the other extreme, if f is constant, $f \circ g = f \circ h$ for all g and h. Provide a specific example.

(b) Suppose, for example, that $f(x) = 73$ for all x and $g(73) = h(73) = 5$ but that $g(x)$ and $h(x)$ are not always equal.

(c) If g and h have the same domain and if f maps onto it, then $g \circ f = h \circ f$ implies $g = h$.

16. (a) $f(f^{-1}(y)) = y \in B$, so $f^{-1}(y) \in f^{\leftarrow}(B)$.

(b) If $x \in f^{\leftarrow}(B)$ then $f(x) \in B$ so $x = f^{-1}(f(x)) \in f^{-1}(B)$.

(c) By (a) and (b).

Section 1.5

1. (a) 42.   (b) 210.   (c) 1.   (d) 1680.

(e) 154.   (f) 360.

2. (a) n.   (b) $n/(n + 1)$.

3. (a) 3, 12, 39 and 120.   (b) 27, 91, 216.   (c) 3, 9 and 45.

4. (a) 0.   (b) 18.   (c) 15.   (d) 10,395.   (e) 2520.

5. (a) -2, 2, 0, 0 and 0.   (b) 2, 3, 4, $m + 1$.

6. (a) 3, 7, 15, 31, 63.   (b) $\sum_{k=0}^{n} 2^k = 2^{n+1} - 1$.

7. (a) 0, 1/3, 1/2, 3/5, 2/3, 5/7.   (b) 1/3, 1/6, 1/10.
   (c) Note that $a_{n+1} = \dfrac{(n+1)-1}{(n+1)+1} = \dfrac{n}{n+2}$ for $n \in \mathbb{P}$. Hence $a_{n+1} - a_n = \dfrac{n}{n+2} - \dfrac{n-1}{n+1} = \dfrac{2}{(n+1)(n+2)}$ for $n \in \mathbb{N}$.

8. (a) 1, 0, 1, 0, 1, 0, 1.   (b) {0,1}.

9. (a) 0, 0, 2, 6, 12, 20, 30.
   (b) Just substitute the values into both sides.
   (c) Same comment.

10. (a) 1, 5, 14, 55.
    (b) By definition of $SSQ(n)$.
    (c) $SSQ(74) = 132,349 + (74)^2 = 137,825$; $SSQ(72) = 132,349 - (73)^2 = 127,020$.

11. (a) 0, 0, 1, 1, 2, 2, 3, 3, ...   (b) 0, 1, 1, 2, 2, 3, 3, ...
    (c) (0,0), (0,1), (1,1), (1,2), (2,2), (2,3), (3,3), ...

12. $\log_2 16 = 4 = \sqrt{16}$; $\log_2 64 = 6 < 8 = \sqrt{64}$; $\log_2 256 = 8 < 16 = \sqrt{256}$; $\log_2 4096 = 12 < 64 = \sqrt{4096}$. As in Example 4(a), it appears that $\log_2 n$ grows more slowly than $\sqrt{n}$.

13. (a)

n	n^4	4^n	n^{20}	20^n	$n!$
5	625	1024	$9.54 \cdot 10^{13}$	$3.2 \cdot 10^6$	120
10	10^4	$1.05 \cdot 10^6$	10^{20}	$1.02 \cdot 10^{13}$	$3.63 \cdot 10^6$
25	$3.91 \cdot 10^5$	$1.13 \cdot 10^{15}$	$9.09 \cdot 10^{27}$	$3.36 \cdot 10^{32}$	$1.55 \cdot 10^{25}$
50	$6.25 \cdot 10^6$	$1.27 \cdot 10^{30}$	$9.54 \cdot 10^{33}$	$1.13 \cdot 10^{65}$	$3.04 \cdot 10^{64}$

(b) The relative growth rates are not obvious from this table. For example, in the long run 4^n will grow faster than n^{20}; in fact, even 2^n will grow faster than n^{20} and n^{200}. These facts will be clarified in the next section.

14. (a)

n	$\log_{10} n$	$\sqrt{n}$	$20 \cdot \sqrt[4]{n}$	$\sqrt[4]{n} \cdot \log_{10} n$
50	1.70	7.07	53.18	4.52
100	2.00	10.00	63.25	6.32
10^4	4.00	100.00	200.00	40.00
10^6	6.00	1000.00	632.46	189.74

(b) As in Exercise 13, the relative growths of these sequences are not clear. $\log_{10} n$ does grow the slowest and $\sqrt{n}$ does grow the fastest. But in the long run $\sqrt[4]{n} \cdot \log_{10} n$ will grow faster than $20 \cdot \sqrt[4]{n}$.

15. $2^n = 10^{n \cdot \log_{10} 2}$. Why? The table values look a little different because the exponents $n \cdot \log_{10} 2$ are not integers.

Section 1.6

1. (a) $k = 2$. (b) $k = 6$ since f is a polynomial of degree 6.
   (c) $k = 12$. (d) $k = \frac{1}{2}$ since $\sqrt{n+1} \le \sqrt{2n} = \sqrt{2} \cdot \sqrt{n}$.

2. (a) $k = 14$. (b) $k = 1$ since $\sqrt{n^2 - 1} \le \sqrt{n^2} = n$.
   (c) $k = 1$. (d) $k = 7$.

3. (a) $n!$ Note that $3^n \ne O(2^n)$ but $3^n = O(n!)$; see Example 3(b) or Exercise 8.
   (b) n^4. (c) $\log_2 n$.

4. (a) n. (b) n^3 since $n \cdot \log_2 n = O(n \cdot \sqrt{n})$.
   (c) n^n. Note that $(n+1)! \ne O(n!)$ since $(n+1)! \le C \cdot n!$ for all large n would imply that $n + 1 \le C$ for all large n.

5. (a) True. Take $C \ge 2$.
   (b) True. $(n+1)^2 = n^2 + 2n + 1$. Apply the result of Example 5(c).
   (c) False. $2^{2n} \le C \cdot 2^n$ only for $2^n \le C$, i.e., only for $n \le \log_2 C$.
   (d) True. Take $C = 40,000$. Or apply Example 5(c).

6. (a) True. By Example 2(c) $\log_2 n = O(n^{1/146})$, so $(\log_2 n)^{73} = O((n^{1/146})^{73}) = O(\sqrt{n})$.
   (b) True because $\log_2(n^{73}) = 73 \cdot \log_2 n$.

13

(c) False because $\log_2 n^n = n \cdot \log_2 n \le C \cdot \log_2 n$ only for $n \le C$.

(d) True because $(\sqrt{n} + 1)^4 \le (2\sqrt{n})^4 = 16 \cdot n^2$.

7. (a) False. If $40^n \le C \cdot 2^n$ for all large n, then $20^n \le C$ for all large n.

(b) True since $(40n)^2 = 1600n^2$.

(c) False. If $(2n)! \le C \cdot n!$ for all large n, then $(n + 1)! \le C \cdot n!$ for large n and so $n + 1 \le C$ for large n, which is impossible.

(d) True because $(n + 1)^{40} \le (2n)^{40} = 2^{40} \cdot n^{40}$. Or apply Example 5(e).

8. For $n > 2A$,
$$n! > n(n - 1) \cdots (2A + 1) > (2A)^{n-2A} = A^n \left\{ 2^n \cdot \frac{1}{(2A)^{2A}} \right\} > A^n$$
for $2^n > (2A)^{2A}$ or $n > 2A \cdot \log_2(2A)$.

9. (a) The inequality $\frac{1}{k^2} \le (\frac{1}{k-1} - \frac{1}{k})$ is equivalent to $(k - 1)k \le k^2$. To check the equality of the hint, write out several terms of the sum.

(b) Clearly $t_n \le n + n + \cdots + n$ [n terms] $= n^2$. In fact, $t_n = \frac{1}{2} \cdot n(n + 1)$ as shown in § 4.2.

10. (a) Set $m = 2$ in the answer to (b).

(b) $t_n = \sum_{k=1}^{n} k^m \le \sum_{k=1}^{n} n^m = n \cdot n^m = n^{m+1}$.

11. We are given $f(n) = 3n^4 + a(n)$ and $g(n) = 2n^3 + b(n)$ where $a(n) = O(n)$ and $b(n) = O(n)$.

(a) Now $f(n) + g(n) = 3n^4 + [a(n) + 2n^3 + b(n)]$ and $a(n) + 2n^3 + b(n) = O(n^3)$ since $a(n)$, $2n^3$ and $b(n)$ are all $O(n^3)$ sequences. Theorem 2(b) with $g(n) = n^3$ is being used here twice.

(b) $f(n) \cdot g(n) = 6n^7 + [2n^3 \cdot a(n) + 3n^4 \cdot b(n) + a(n) \cdot b(n)]$. Theorem 2(c) shows that $2n^3 \cdot a(n) = O(n^4)$, $3n^4 \cdot b(n) = O(n^5)$ and $a(n) \cdot b(n) = O(n^2)$. So the sum in brackets is $O(n^5)$ by Theorem 2(b) with $g(n) = n^5$. Or apply Theorem 3(a).

12. (a) $5n^3 + O(n^2)$ stands for $5n^3 + a(n)$ where $a(n) = O(n^2)$. Similarly, $3n^4 + O(n^3)$ stands for $3n^4 + b(n)$ with $b(n) = O(n^3)$. Now $(5n^3 + a(n))(3n^4 + b(n)) = 15n^7 + [a(n) \cdot 3n^4 + 5n^3 \cdot b(n) + a(n) \cdot b(n)]$. By Theorem 2(c), $a(n) \cdot 3n^4 = O(n^6)$, $5n^3 \cdot b(n) = O(n^6)$ and $a(n) \cdot b(n) = O(n^5)$, so the sum in brackets is $O(n^6)$.

(b) The argument is similar to that for (a).

13. (a) Assume $c \neq 0$ since the result is obvious for $c = 0$. There is a $C > 0$ so that $|f(n)| \leq C \cdot |g(n)|$ for large n. Then $|c \cdot f(n)| \leq C \cdot |c| \cdot |g(n)|$ for large n.

(c) If $|f(n)| \leq C \cdot |a(n)|$ and $|g(n)| \leq D \cdot |b(n)|$ for large n, then $|f(n) \cdot g(n)| \leq C \cdot D \cdot |a(n) \cdot b(n)|$ for large n.

14. (a) If $f(n) = O(a(n))$ and $g(n) = O(b(n))$, then $f(n) = O(c(n))$ and $g(n) = O(c(n))$ by Theorem 2(d). Apply Theorem 2(b).

(b) To get Theorem 3(a), take $c(n) = O(\max\{|a(n)|, |b(n)|\})$.

15. (a) $a(n)/b(n) = n^4$ is not $O(n^3)$, by Example 8(e).

(b) For example, $a(n) = n^6$ and $b(n) = n$.

16. Since $\log_{10} n = \log_{10} 2 \cdot \log_2 n$ and $\log_{10} 2$ is a constant, this follows from Theorem 2(a).

17. (a) Let $\mathrm{DIGIT}(n) = m$. Then $10^{\mathrm{DIGIT}(n)}$ is written as a 1 followed by m 0's, so it's larger than any m-digit number, such as n. And 10^{m-1} is a 1 followed by m - 1 0's so it's the smallest m-digit number.

(b) Since $n < 10^{\mathrm{DIGIT}(n)}$ by part (a),
$$\log_{10} n < \log_{10} 10^{\mathrm{DIGIT}(n)} = \mathrm{DIGIT}(n) \quad \text{for every } n \in \mathbb{P}.$$

(c) Use part (a). In detail, $\mathrm{DIGIT}(n) - 1 = \log_{10} 10^{\mathrm{DIGIT}(n) - 1} \leq \log_{10} n$, so $\mathrm{DIGIT}(n) \leq 1 + \log_{10} n = O(\log_{10} n)$.

(d) They mean the same thing. Both are equivalent to $O(\log_2 n)$.

CHAPTER 2

This chapter introduces logic, both formal logic with its truth tables, tautologies and formal proofs, and logical thinking and proof methods as they are used in practice to give convincing arguments. Truth table methods come up again in the Boolean setting of Chapter 10. Logical thinking is expected from this point on, no matter what the context.

Section 2.1 is a gentle introduction to logical terms and symbols; § 2.2 gets down to business, with truth tables for the connectives and for compound propositions. Students seem to have special trouble with $p \rightarrow q$ in the cases in which p is false. One possible justification, which we do not give in the text, is that the other ways those entries could be filled in correspond to the legitimate propositions $p \wedge q$, q and $p \leftrightarrow q$, whose meanings are different from what $p \rightarrow q$ should mean. Give lots of examples.

Don't let students get hung up on Tables 1 and 2 in § 2.2. They should be able to use the tables, but don't need to memorize them. Discuss Example 7 of § 2.2, and explain why some rows of the truth table can be ignored.

Section 2.3 is important. Not all students will pick these ideas up the first time. Encourage students to refer back to this section as the need arises. Example 10 slips in an infinite-finite pigeonhole principle without fanfare and then applies it to looking at remainders. The application foreshadows modular arithmetic in § 3.6.

The substitution rules in § 2.4 just formalize two obvious ways of getting new equivalences from old. To paraphrase: (b) says you can always replace a proposition by an equivalent one, while (a) says you can replace a variable p by a proposition in a tautology provided you replace it with the same proposition everywhere.

Students [and instructors] often wonder why we spend time on formal proofs, with their ritualistic format. Our main concern is to develop logical reasoning, not to build facility with formal proofs. We include the formalism to illustrate that there is a bedrock not far below the level at which we normally work. In previous editions we scared some students by asking them to construct formal proofs. In this edition we just ask them to supply reasons; see Exercises 13 and 14. Exercises 16-18 have applications in circuit design. We do one of the three in class and assign one or both of the others as homework.

Section 2.5 begins with examples of translation between "real life" logic and formal proofs. Its main purpose is to give students practical guidance for constructing acceptable proofs and recognizing gaps and fallacies. This section is quite informal in spirit. You may want to draw examples from your own experience or get the class to volunteer sample arguments. Some of the exercises do ask for formal proofs, but the constructions are not complicated.

Section 2.1

1. (a) $p \wedge q$.   (b) $p \rightarrow r$.   (c) $\neg p \rightarrow (\neg q \wedge r)$.   (d) $q \leftrightarrow (\neg p)$.
   (e) $\neg r \rightarrow q$.

2. (a) If it is raining and the sun is shining, then there are clouds in the sky.
   (b) If it is raining only if there are clouds in the sky, then the sun is shining.
   (c) It is not raining if and only if [either] the sun is shining or there are clouds in the sky.
   (d) It is not true that it is raining if and only if [either] the sun is shining or there are clouds in the sky.
   (e) It is not true that it is raining or the sun is shining, and there are clouds in the sky.
   There are, of course, other ways to express these propositions in English.

3. (a) Parts (b) and (c) are true. The other three are false.
   (b) In Example 2, parts (a) and (b) are true.

4. (a) False.   (b) True.   (c) Not a well-defined proposition.
   (d) False.   (e) Not a proposition.   (f) False; x might be 0.   (g) True.

5. The proposition is true for all $x, y \in [0, \infty)$, but is false when applied to all $x, y \in \mathbb{R}$.

6. (a) $r \rightarrow q$.
   (b) If I am rich, then I am smart.
   (c) If $x = 0$ or $x = 1$, then $x^2 = x$.
   (d) If $2 + 4 = 8$, then $2 + 2 = 4$.

7. (a) $\neg r \rightarrow \neg q$.   (b) If I am not rich, then I am not smart.
   (c) If it is false that $x = 0$ or $x = 1$, then $x^2 \neq x$.
   (d) If $2 + 4 \neq 8$, then $2 + 2 \neq 4$.

8. (a) $6 = 3 + 3, 8 = 3 + 5, 10 = 3 + 7$.   (b) $98 = 19 + 79$.

9. (a) $3^3 < 3^3$ is false. (b) There are no other counterexamples.

10. (b) Any pair $(n,-n)$ will do, $n \neq 0$.

11. (a) $(-1 + 1)^2 = 0 < 1 = (-1)^2$. (b) choose $x < -\frac{1}{2}$.
    (c) No. If $x \geq 0$, then $(x + 1)^2 = x^2 + 2x + 1 > x^2$. In fact, if $x \geq -\frac{1}{2}$, then $2x + 1 \geq 0$
    and $(x + 1)^2 \geq x^2$.

12. (a) $n = 6$. (b) $n = 3$. (c) $n = 7$.

13. (a) $(0,-1)$. (b) Restrict x, y to be nonnegative.

14. (a) True by a commutative law for sets, Table 1 of § 1.2.
    (b) False. choose A and B with B not contained in A. For example, choose
    $A = \emptyset \neq B$.
    (c) False. choose A and B to overlap. For example, $A = B \neq \emptyset$.
    (d) True by an associative law for sets.

15. (a) $p \to q$. (b) $p \to r$. (c) $\neg r \to p$. (d) $q \to p$. (e) $r \to q$.
    (f) $r \to (q \vee p)$ or $(r \to q) \vee p$. Punctuation would help in part (f). For example, a
    comma after "$I = 0$" would yield $(r \to q) \vee p$.

16. (a) $r \to m$. (b) $\neg r \to p$. (c) $r \to m$. (d) $p \to m$.
    (e) $r \vee (p \to m)$ or $p \to (r \vee m)$. Again punctuation would help.

17. (a) The probable intent is "If you touch those cookies, then I will spank you." It is easier
    to imagine p of $p \to q$ being true for this meaning than it is for "If you want a spanking,
    then touch those cookies."
    (b) If you touch those cookies, then you will be sorry.
    (c) If you do not leave, then I will set the dog on you.
    (d) If you will, then I will.
    (e) If you do not stop that, then I will go.

18. (a) "If I will not spank you, then you did not touch those cookies." In the other version,
    the contrapositive is "If you do not touch those cookies, then you do not want a spanking."
    (b) If you will not be sorry, then you will not touch those cookies.
    (c) If I will not set the dog on you, then you do leave.

(d) If I won't, then you won't.

(e) If I will not go, then you do stop that.

Section 2.2

1. (a) Converse: $(q \wedge r) \rightarrow p$.
   Contrapositive: $\neg (q \wedge r) \rightarrow \neg p$.

   (b) Converse: If $x^2 + y^2 \geq 1$, then $x + y = 1$.
   Contrapositive: If $x^2 + y^2 < 1$, then $x + y \neq 1$.

   (c) Converse: If $3 + 3 = 8$, then $2 + 2 = 4$.
   Contrapositive: If $3 + 3 \neq 8$, then $2 + 2 \neq 4$.

2. (a) Converse: If $x^2 > 0$, then $x > 0$.
   Contrapositive: If $x^2 \leq 0$, then $x \leq 0$.

   (b) The proposition and its contrapositive are true. The converse is false.

3. (a) $q \rightarrow p$.　　(b) $\neg q \rightarrow \neg p$.　　(c) $p \rightarrow q$, $\neg q \rightarrow \neg p$, $\neg p \vee q$.

4. All but (a) and (g) are true.

5. (a) 0.　　(b) 1.　　(c) 1.

Note. For some truth tables, only the final columns are given.

6.

p	part(a)	part(b)	part(c)	part(d)
0	0	1	0	0
1	0	1	0	1

7.

p	q	part (a)	part (b)	part(c)	part (d)
0	0	1	1	1	1
0	1	1	0	0	1
1	0	1	0	0	1
1	1	0	0	0	0

8.

p	q	(p→q)	→	[(p ∨ ¬q)	→	(p ∧ q)]
0	0	1	0	1 1	0	0
0	1	1	1	0 0	1	0
1	0	0	1	1 1	0	0
1	1	1	1	1 0	1	1

9.

p	q	r	final column
0	0	0	1
0	0	1	1
0	1	0	1
0	1	1	0
1	0	0	1
1	0	1	1
1	1	0	1
1	1	1	0

10.

p	q	r	final column
0	0	0	0
0	0	1	0
0	1	0	0
0	1	1	0
1	0	0	1
1	0	1	1
1	1	0	0
1	1	1	0

11.

p	q	r	part (a)	part (b)
0	0	0	0	0
0	0	1	1	0
0	1	0	1	1
0	1	1	1	0
1	0	0	1	1
1	0	1	1	0
1	1	0	1	1
1	1	1	1	0

12. (a) Exclusive; only one of soup or salad is offered.

   (b) Inclusive; having both would be fine.

   (c) The original intent was exclusive, but cynics might say it's inclusive.

   (d) Same as (b).

   (e) Exclusive; it can't be completed on both days.

   (f) Exclusive.

   (g) The intention is ¬f ∧ ¬h ⇔ ¬(f ∨ h), so ∨ is inclusive.

   (h) ¬J ∧ ¬A ⇔ ¬(J ∨ A), so inclusive.

13. (b)

(d)

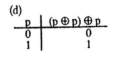

(c)

p	q	r	(p $\oplus$ q) $\oplus$ r
0	0	0	0
0	0	1	1
0	1	0	1
0	1	1	0
1	0	0	1
1	0	1	0
1	1	0	0
1	1	1	1

14. (a) $\{[(p \wedge \neg q) \wedge \neg r] \vee [(\neg p \wedge q) \wedge \neg r]\} \vee [(\neg p \wedge \neg q) \wedge r]$ is one such proposition. So is $[(p \leftrightarrow \neg q) \wedge \neg r] \vee [\neg (p \vee q) \wedge r]$.

(b) $\{[(p \wedge q) \wedge \neg r] \vee [(p \wedge \neg q) \wedge r]\} \vee [(\neg p \wedge q) \wedge r]$ is one. So is $[p \wedge (q \leftrightarrow \neg r)] \vee [\neg p \wedge (q \wedge r)]$.

15. (a) No fishing is allowed and no hunting is allowed. The school will not be open in July and the school will not be open in August.

(b) No. $\neg (p \oplus q) \Leftrightarrow [(p \wedge q) \vee (\neg p \wedge \neg q)]$. For example, to negate Exercise 12(a) one would need something like "You must choose both soup and salad or else neither soup nor salad."

16. (a) True. Both $p \to (q \to r)$ and $(p \to q) \to (p \to r)$ are false only for

p	q	r
1	1	0

(b) False. consider

p	q	r
1	0	0

or

p	q	r
1	1	1

(c) The propositions are not equivalent, as can be verified by comparing the first or third rows of their truth tables.

(d) and (e) are true. It is probably easiest to verify them using truth tables.

19. (a) One need only consider rows in which $[(p \wedge r) \to (q \wedge r)]$ is false, i.e., $(p \wedge r)$ is true and $(q \wedge r)$ is false. This leaves one row to consider:

p	q	r
1	0	1

(b) It's easiest to show that $p \to q$ has truth value 0 whenever $(q \to r) \to (p \to r)$ has truth value 0. Now $(q \to r) \to (p \to r)$ is false precisely if $q \to r$ is true and $p \to r$ is false, i.e., if p is true, r is false and q is false, in which case $p \to q$ is false, as desired. That is, the only row that matters is

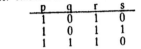

(c) One need only consider rows in which $[(p \wedge r) \rightarrow (q \wedge s)]$ is false, i.e., $(p \wedge r)$ is true and $(q \wedge s)$ is false. This leaves three rows to consider:

p	q	r	s
1	0	1	0
1	0	1	1
1	1	1	0

20. (a) False. consider $q = 0$.

(b) False. consider $p = 1$, $q = 0$.

(c)

p	q	$(p \wedge q)$	$\rightarrow$	$(p \vee q)$
0	0	0	1	0
0	1	0	1	1
1	0	0	1	1
1	1	1	1	1

Thus $(p \wedge q) \rightarrow (p \vee q)$ is a tautology, so $(p \wedge q) \Rightarrow (p \vee q)$ is true.

21. Let $p =$ "He finished dinner" and $q =$ "He was sent to bed." Then p is true and q is true, so the logician's statement $\neg p \rightarrow \neg q$ has truth value True. She was logically correct, but not very nice.

22. (a) If concrete grows, then you do water it. [The original statement is "If you do not water concrete, then it does not grow."]

(b) If concrete does not grow, then you do not water it.

(c) If you water concrete, then it grows.

(d) The original and its contrapositive are true. The statements in (b) and (c) are false.

23. (a) Consider the truth tables. B has truth value 1 on every row that A does, and C has truth value 1 on every row that B does, so C has truth value 1 on every row that A does.

(b) Since $P \Leftrightarrow Q$, $P \Rightarrow Q$. By (a), $P \Rightarrow R$. Since $R \Leftrightarrow S$, $R \Rightarrow S$. By (a), $P \Rightarrow S$. Or consider truth tables.

(c) We are given that $P \Rightarrow Q$. Since $Q \Rightarrow R$ and $R \Rightarrow P$, by (a) $Q \Rightarrow P$. Thus $P \Leftrightarrow Q$.

Section 2.3

1. Give a direct proof using the following fact. If m and n are even integers, then there exist j and k in $\mathbb{Z}$ so that $m = 2j$ and $n = 2k$.

2. The even integer is $2j$ and the odd integer is $2k + 1$ for some j and k in $\mathbb{Z}$. Their product $2j \cdot (2k + 1) = 2[j \cdot (2k + 1)]$ is even. [Indeed, it's irrelevant that the second integer is odd.] This proof is direct.

3. This can be done using four cases: see Example 5.

4. Factor $n^4 - n^2$ into $(n - 1)n^2(n + 1)$ and note that one of the three consecutive integers $n - 1, n, n + 1$ is divisible by 3. Or treat three cases: $n = 3k$, $n = 3k + 1$, $n = 3k + 2$. Either proof is direct.

5. This can be done using three cases: (i) $n = 3k$ for some $k \in \mathbb{N}$; (ii) $n = 3k + 1$ for some $k \in \mathbb{N}$; (iii) $n = 3k + 2$ for some $k \in \mathbb{N}$.

6. (a) Imitate the proof in Example 2, using an analogue of Example 4 to show, by considering cases, that if p is not divisible by 3 then p^2 is not divisible by 3. The Example 2 proof is by contradiction; the Example 4 proof is direct. [See also the answer to part (b).]

(b) Imitate the proofs in Examples 2 and 4. Assume $x = m/n$ with $m \in \mathbb{Z}$, $0 \neq n \in \mathbb{Z}$ and $x^3 = 2$. We may suppose m and n have no common factors. Then $m^3 = 2n^3$. Since $(2k + 1)^3 = 8k^3 + 12k^2 + 6k + 1$ is odd, m cannot be odd. I.e., $m^3 = 2n^3$ and m odd lead to a contradiction. So m is even, say $m = 2k$. Then $8k^3 = 2n^3$, so $4k^3 = n^3$. By the same reasoning, n must be even. But then m and n have a common factor, namely 2, and this is a contradiction.

This proof is by contradiction.

7. (a) Give a direct proof, as in Exercise 1.

(b) False. The sum of two odd integers is **always** even. For instance, $1 + 1 = 2$.

(c) False. $2 + 3$ or $2 + 5$ or $2 + 11$, for instance.

(d) True. Say the integers are $n, n + 1$ and $n + 2$. Their sum is $3n + 3 = 3(n + 1)$.

(e) False. Finding an example will be easy; the sum of four consecutive integers is **never** divisible by 4.

(f) True; $n + (n + 1) + (n + 2) + (n + 3) + (n + 4) = 5n + 10$. Alternatively, $(n - 2) + (n - 1) + n + (n + 1) + (n + 2) = 5n$.

8. (a) Some examples: $(17,19)$, $(29,31)$, $(41,43)$, $(1997,1999)$.

(b) Consider three cases: $k = 3m$, $k = 3m + 1$ and $k = 3m + 2$ [or $3m - 1$]. For example, if $k = 3m$, then $2k + 3 = 6m + 3 = 3(2m + 1)$ is divisible by 3.

9. (a) Trivially true. (b) True, because $p \rightarrow p$ is a tautology. (c) Vacuously true.

10. Suppose $xy = 0$. Then either $x = 0$ or $y = 0$. Say $x = 0$. Then $x^n = 0^n = 0$ and $(x + y)^n = (0 + y)^n = y^n = 0 + y^n = x^n + y^n$. The argument for $y = 0$ is similar.

11. Example 2 shows that the set of primes is infinite. An argument like the one in Example 10 shows that some two primes give the same last six digits. This proof is nonconstructive.

12. The first few values of a_n are positive, by inspection. According to the hint, all of the values of a_n are negative for n large enough, so a_n only has finitely many positive values. One of them must be as large as they get. [Perhaps more than one, though we doubt it.] This proof is not constructive. Computing the values of a_n is not a very practical way to proceed either, though it could be done.

13. (a) None of the numbers in the set
$$\{k \in \mathbb{N} : (n + 1)! + 2 \leq k \leq (n + 1)! + (n + 1)\}$$
is prime, since if $2 \leq m \leq n + 1$, then m divides $(n + 1)!$ and so m also divides $(n + 1)! + m$.
(b) Yes. Since $7! = 5040$, the proof shows that all the numbers from 5042 to 5047 are nonprime.
(c) Simply adjoin 5048 to the list in part (b). Another sequence of seven nonprimes starts with 90.

14. Each p_i divides $p_1 p_2 \cdots p_k$, so it can't divide $1 + p_1 p_2 \cdots p_k$. [If it did, it would divide their difference, which is 1.] So every prime factor of $1 + p_1 p_2 \cdots p_k$ must be different from $p_1, p_2, \ldots, p_k$. [At this writing, the problem of factoring an arbitrary integer is conjectured to be NP-hard. Thus there may be no effective algorithm for carrying out the method of this exercise to construct new large primes.]

15. (a) $14 = 2 \cdot 7$ and 7 is odd. So $14 = 2^1 \cdot 7$.
(b) 73 is odd, so $73 = 2^0 \cdot 73$.
(c) $96 = 2 \cdot 48 = 2 \cdot 2 \cdot 24 = 2 \cdot 2 \cdot 2 \cdot 12 = 2 \cdot 2 \cdot 2 \cdot 2 \cdot 6 = 2 \cdot 2 \cdot 2 \cdot 2 \cdot 2 \cdot 3$, so $96 = 2^5 \cdot 3$.
(d) Answer $= 2^4 \cdot 73$.

16. (a) $2^4 - 1 = 15$ works.

(b) This is already a lot harder, but the last paragraph of Example 10 suggests trying various numbers $2^n - 1$. One finds $2^{10} - 1 = 1023 = 93 \cdot 11$. [There is a general principle at work here; can you guess it?]

Section 2.4

1. (a) $\neg (p \rightarrow q) \rightarrow ((p \rightarrow q) \rightarrow p)$

(b) $[p \wedge (p \rightarrow (p \rightarrow q))] \rightarrow (p \rightarrow q)$

(c) $p \vee \neg p$

(d) $[(p \vee (p \rightarrow q)] \leftrightarrow [\neg (p \rightarrow q) \rightarrow p]$

2. For example, rule 23 corresponds to the rule of inference

$$P \leftrightarrow Q$$
$$\underline{Q \leftrightarrow R}$$
$$\therefore \ P \leftrightarrow R$$

26a corresponds to

$$P \rightarrow Q$$
$$\underline{R \rightarrow S}$$
$$\therefore \ (P \vee R) \rightarrow (Q \vee S)$$

27b corresponds to

$$P \rightarrow Q$$
$$\underline{R \rightarrow S}$$
$$\therefore \ (\neg Q \wedge \neg S) \rightarrow (\neg P \wedge \neg R)$$

3. (a) Rule 14 with $q \wedge r$ replacing q.

(b) Rule 4a with $r \wedge s$ replacing r.

(c) Rule 8a with $\neg p \wedge r$ replacing p and $q \rightarrow r$ replacing q.

4. (a) Rule 22 with $\neg p \vee q$ replacing p.

(b) Rule 27a with q replacing s.

(c) Rule 20 with $p \rightarrow s$ replacing p and $q \wedge s$ replacing q.

5. (a) Rule 2a and Substitution Rule (b).

(b) Rule 8a and Rule (b).

(c) Rules 10a and 1 and Rule (b).

(d) First apply Rule (a) to rule 2a to get $s \vee p \Leftrightarrow p \vee s$. Then use Rule (b).

6. (a) Rule 5b and Substitution Rule (b).

    (b) Rule 10a and Rule (b).

    (c) Rules 2a [with a for p, b for q, using Substitution Rule (a)] and 8b, and Rule (b).

7. (a) Rule 10a [with s for q, using Rule (a)] and Rule 11a [with s for p and t for q, using Rule (a)] and Rule (b).

    (b) Rule 3a [with $\neg p \vee s$ for p, s for q, t for r, using Rule (a)].

    (c) Rule 3a [with $\neg p$ for p, s for q, s for r, using Rule (a)] and Rule (b).

    (d) Rule 5a [with s for p, using Rule (a)] and Rule (b).

    (e) Rule 3a again, using Rule (a).

    (f) Rule 10 [with $s \vee t$ for q, using Rule (a)].

8. (a) Rule 10a and Rule (a).

    (b) Rule 8a and Rules (a) and (b).

    (c) Rule 8b [with Rule (a)] and Rule (b).

    (d) Rule 2a with Rule (a).

    (e) Rule 4a with Rule (a).

    (f) Rules 2a [with Rule (a)] and 7a and Rule (b).

    (g) Rule 6d with Rule (a).

    (h) Rule 3a with Rule (a).

    (i) Rules 2a and 7a and Rule (b).

    (j) Rule 6b with Rule (a).

9. We obtain successive equivalences using the indicated rules and suitable substitutions.

    (a)

$[(p \vee r) \wedge (q \rightarrow r)]$	
$[(p \vee r) \wedge (\neg q \vee r)]$	Rule 10a
$[(r \vee p) \wedge (r \vee \neg q)]$	Rule 2a twice
$r \vee (p \wedge \neg q)$	Rule 4a
$(p \wedge \neg q) \vee r$	Rule 2a

    (b)

$(p \wedge \neg q) \vee r$	
$\neg(\neg p \vee \neg \neg q) \vee r$	Rule 8d
$\neg(\neg p \vee q) \vee r$	Rule 1
$\neg(p \rightarrow q) \vee r$	Rule 10a
$(p \rightarrow q) \rightarrow r$	Rule 10a

    (c)

$p \vee (\neg p \wedge \neg q)$	
$(p \vee \neg p) \wedge (p \vee \neg q)$	Rule 4a
$t \wedge (p \vee \neg q)$	Rule 7a

$p \vee \neg q$ Rules 2b and 6d

10. (a) Rule 16.

    (b) $(p \vee q) \rightarrow p \Leftrightarrow [\neg(p \vee q) \vee p]$ Rule 10a [and Rule (b)]

                     $\Leftrightarrow [(\neg p \wedge \neg q) \vee p]$ Rule 8a [and Rule (b)]

                     $\Leftrightarrow [p \vee (\neg p \wedge \neg q)]$ Rule 2a [and Rule (b)]

    (c) $[p \vee (\neg p \wedge \neg q)] \Leftrightarrow (p \vee \neg q)$ Exercise 9(c)

                     $\Leftrightarrow \neg q \vee p$ Rule 2a

    (d) By parts (a), (c) and (b).

11. Consider the cases $\underset{1\ 0\ 1}{p\ q\ r}$ or $\underset{1\ 1\ 0}{p\ q\ r}$, and $\underset{0\ 0\ 1}{p\ q\ r}$ or $\underset{0\ 1\ 0}{p\ q\ r}$.

12. Consider the row of the truth table where p is false and q is true.

13. 1,2,3: Hypothesis

     4: Rule 16, a tautology

     5: 4, 1 and hypothetical syllogism (rule 33)

     6: 5, 3 and rule 33

     7: 6, 2 and modus tollens (rule 31)

14. All that matters is that 1 and 4 precede 5, 3 and 5 precede 6, and 2 and 6 precede 7. One possible order is:

    1. $s \vee g \rightarrow p$

    2. $s \rightarrow s \vee g$ [old 4]

    3. $s \rightarrow p$ [old 5]

    4. $p \rightarrow a$ [old 3]

    5. $s \rightarrow a$ [old 6]

    6. $\neg a$ [old 2]

    7. $\neg s$

15. (a) Take the original proof and change the reason for A from "hypothesis" to "tautology." I.e., the proof itself needs no change.

    (b) Let A be a tautology in Example 8, and use (a).

    Specifically, if there is a proof of C from B, then there is a proof of C from $A \wedge B$, so by Example 8 there is a proof of $B \rightarrow C$ from A. Apply part (a).

16. (a) $\neg(\neg p \vee \neg q) \vee \neg(p \vee q)$.

    (b) $(\neg p \vee \neg q) \vee \neg(q \vee \neg r)$.

    (c) $\neg[\neg(\neg p \vee q) \vee \neg(q \vee r)]$.

    (d) $(p \wedge \neg q) \vee (\neg p \wedge q)$ leads to $\neg(\neg p \vee q) \vee \neg(p \vee \neg q)$.

    $(p \vee q) \wedge \neg(p \wedge q)$ leads to $\neg[\neg(p \vee q) \vee \neg(\neg p \vee \neg q)]$.

17. (a) See rules 11a and 11b.

    (b) $p \vee q \Leftrightarrow \neg(\neg p \wedge \neg q)$, $\quad p \rightarrow q \Leftrightarrow \neg(p \wedge \neg q)$.

    (c) No. Any proposition involving only $p, q, \wedge$ and $\vee$ will have truth value 0 whenever p and q both have truth values 0. See Exercise 17 of § 7.2.

18.

p q	(p|p)	|	(q|q)
0 0	1	0	1
0 1	1	1	0
1 0	0	1 .	1
1 1	0	1	0

gives (a) and (b).

   (c) Since $p\|q \Leftrightarrow \neg(p \wedge q)$ by inspection, $p \wedge q \Leftrightarrow \neg(p\|q) \Leftrightarrow (p\|q)\|(p\|q)$, by part (a).

   (d) $p \rightarrow q \Leftrightarrow (\neg p) \vee q \Leftrightarrow \neg(p \wedge (\neg q)) \Leftrightarrow p\|(\neg q) \Leftrightarrow p\|(q\|q)$.

   (e) $p \oplus q$ can be written out in very long form by using the results of parts (a) and (b). The shorter answer $[p\|(q\|q)]\|[q\|(p\|p)]$ comes from part (d) and the observation that $p \oplus q \Leftrightarrow \neg[(p \rightarrow q) \wedge (q \rightarrow p)]$; see Exercise 13 of § 2.2.

19. Use truth tables.

Section 2.5

1. The argument is not valid, since the hypotheses are true if C is true and A is false. The error is in treating $A \rightarrow C$ and $\neg A \rightarrow \neg C$ as if they were equivalent.

3. (a) and (b). See (c).

   (c) The case is no stronger. If C and all A_i's are false then every hypothesis $A_i \rightarrow C$ is true, whether or not C is true.

4. (a) No. B could be true but M, D and S false.

   (b) Same answer as (a).

5. (a) With suggestive notation, the hypotheses are $\neg b \rightarrow \neg s$, $s \rightarrow p$ and $\neg p$. We can infer $\neg s$ using the contrapositive. We cannot infer either b or $\neg b$. Of course, we can infer more complex propositions, like $\neg p \vee s$ or $(s \wedge b) \rightarrow p$.

(b) I did not pass both the midterm and the final. I.e., I failed at least one of them.

(c) The hypotheses are $(m \vee f) \rightarrow c$, $n \rightarrow c$ and $\neg n$. No interesting conclusions, such as m or $\neg c$, can be inferred.

6. (a) True if the only possibilities are bus, subway and cab, since of these three only the cab would make me on time. If there are other possibilities, a cab cannot be inferred.

(b) The answer is similar to that for part (a).

(c) Must follow. Since $b \vee s \rightarrow \ell$ and $\neg \ell$, $\neg (b \vee s)$ by modus tollens, so $\neg b \wedge \neg s$.

(d) The answer is similar to that for part (a). If a cab was taken, then the implication is true whether I become broke or not. If a cab was not taken, then the implication is true if and only if I didn't become broke.

(e) Must follow. From (c)'s answer and simplification (rule 29), $\neg b$ can be inferred. By addition (rule 28), $\neg b \vee n$ can be inferred, where n means "not broke." So $b \rightarrow n$ by implication (rule 10a). In fact, $b \rightarrow p$ can be inferred for any p.

7. (a) True. $A \rightarrow B$ is a hypothesis. We showed that $\neg A \rightarrow \neg B$ follows from the hypotheses.

(b) False. See (d) or check the case A, B, Y, L, N all true.

(c) True. Since $(B \vee \neg Y) \rightarrow A$ is given, $\neg Y \rightarrow A$ follows, or equivalently $\neg \neg Y \vee A$.

(d) True. By hypothesis $(B \vee Y) \rightarrow (L \wedge N)$. By (a) and substitution we have $(A \vee Y) \rightarrow (L \wedge N)$. Apply (c).

8. Pat did it. Here is the record from the trial, showing P from $P \vee Q$, $\neg (R \wedge Q)$ and R.

1.	$\neg (R \wedge Q)$	hypothesis
2.	$(\neg R) \vee (\neg Q)$	1; rule 8b
3.	R	hypothesis
4.	$\neg \neg R$	3; rule 1
5.	$\neg Q$	2,4; disjunctive syllogism rule 32
6.	$P \vee Q$	hypothesis
7.	$Q \vee P$	6; rule 2a (commutative law)
8.	P	7,5; rule 32

9. (a) Let $c =$ "my computations are correct," $b =$ "I pay the electric bill," $r =$ "I run out of money," and $p =$ "the power stays on." Then the theorem is:

if $(c \wedge b) \rightarrow r$ and $\neg b \rightarrow \neg p$, then $(\neg r \wedge p) \rightarrow \neg c$.

1.	$(c \wedge b) \rightarrow r$	hypothesis
2.	$\neg b \rightarrow \neg p$	hypothesis

29

3. $\neg r \rightarrow \neg (c \wedge b)$ 1; contrapositive rule 9
4. $\neg r \rightarrow (\neg c \vee \neg b)$ 3; DeMorgan law 8b
5. $p \rightarrow b$ 2; contrapositive rule 9
6. $(\neg r \wedge p) \rightarrow [(\neg c \vee \neg b) \wedge b]$ 4,5; rule of inference corresponding to rule 26b
7. $(\neg r \wedge p) \rightarrow [b \wedge (\neg c \vee \neg b)]$ 6; commutative law 2b
8. $(\neg r \wedge p) \rightarrow [(b \wedge \neg c) \vee (b \wedge \neg b)]$ 7; distributive law 4b
9. $(\neg r \wedge p)$
   $\rightarrow [(b \wedge \neg c) \vee \text{contradiction}]$ 8; rule 7b
10. $(\neg r \wedge p) \rightarrow (b \wedge \neg c)$ 9; identity law 6a
11. $(\neg r \wedge p) \rightarrow (\neg c \wedge b)$ 10; commutative law 2b
12. $(\neg c \wedge b) \rightarrow \neg c$ simplification (rule 17)
13. $(\neg r \wedge p) \rightarrow \neg c$ 11,12; hypothetical syllogism

(b) If $d \rightarrow (h \vee s)$ and $s \leftrightarrow w$, then $\neg h \rightarrow (\neg d \vee w)$.

1. $d \rightarrow (h \vee s)$ hypothesis
2. $s \leftrightarrow w$ hypothesis
3. $d \rightarrow (s \vee h)$ 1; commutative law 2a
4. $(s \rightarrow w) \wedge (w \rightarrow s)$ 2; equivalence rule 13
5. $s \rightarrow w$ 4; simplification (rule 29)
6. $(s \vee h) \rightarrow (w \vee h)$ 5; rule of inference based on rule 25a
7. $d \rightarrow (w \vee h)$ 3,6; hypothetical syllogism (rule 33)
8. $(\neg d) \vee (w \vee h)$ 7; implication rule 10a
9. $(\neg d \vee w) \vee h$ 8; associative law 3a
10. $h \vee (\neg d \vee w)$ 9; commutative law 2a
11. $\neg (\neg h) \vee (\neg d \vee w)$ 10; double negation rule 1
12. $\neg h \rightarrow (\neg d \vee w)$ 11; implication rule 10a

(c) Let j = "I get the job," w = "I work hard," p = "I get promoted," and h = "I will be happy." Then the theorem is: if $(j \wedge w) \rightarrow p$, $p \rightarrow h$ and $\neg h$, then $\neg j \vee \neg w$.

1. $(j \wedge w) \rightarrow p$ hypothesis
2. $p \rightarrow h$ hypothesis
3. $\neg h$ hypothesis
4. $\neg p$ 2,3; modus tollens (rule 31)
5. $\neg (j \wedge w)$ 1,4; modus tollens (rule 31)
6. $\neg j \vee \neg w$ 5; DeMorgan law 8b

(d) If $\ell \rightarrow m$, $a \rightarrow t$ and $(m \vee t) \rightarrow \neg d$, then $d \rightarrow (\neg \ell \wedge \neg a)$.

1. $\ell \rightarrow m$ hypothesis
2. $a \rightarrow t$ hypothesis
3. $(m \vee t) \rightarrow \neg d$ hypothesis
4. $(\ell \vee a) \rightarrow (m \vee t)$ 1,2; rule of inference corresponding to rule 26a
5. $(\ell \vee a) \rightarrow \neg d$ 3,4; hypothetical syllogism (rule 33)
6. $\neg (\neg d) \rightarrow \neg (\ell \vee a)$ 5; contrapositive rule 9
7. $d \rightarrow \neg (\ell \vee a)$ 6; double negation rule 1
8. $d \rightarrow (\neg \ell \wedge \neg a)$ 7; DeMorgan law 8a

10. (a) Replace lines 8, 9 and 10 with:

8. $\neg$ n	hypothesis
9. $\neg$ s	5,8; modus tollens rule 31
10. s $\wedge \neg$ s	7,9; rule 34

(b) Replace as follows:

8. p	3,7; modus ponens rule 30
9. $\neg$ n	hypothesis
10. $\neg$ p	4,9; modus tollens rule 31
11. p $\wedge \neg$ p	8,10; rule 34

11.

1. $\neg \neg$ s	negation of conclusion
2. s	1; rule 1
3. s $\rightarrow$ s $\vee$ g	addition
4. s $\vee$ g	2,3; modus ponens
5. s $\vee$ g $\rightarrow$ p	hypothesis
6. p	4,5; modus ponens
7. p $\rightarrow$ n	hypothesis
8. n	6,7; modus ponens
9. $\neg$ n	hypothesis
10. n $\wedge \neg$ n	8,9; conjunction
11. contradiction	10

12. (a) The theorem is false. Consider (p,q,r) in the set {(0,0,0), (0,0,1), (0,1,0)}.

(b) The theorem is false; suppose p is false and either q is true or r is false.

(c)

1. p $\rightarrow$ (q $\vee$ r)	hypothesis
2. q $\rightarrow$ s	hypothesis
3. r $\rightarrow \neg$ p	hypothesis
4. (q $\vee$ r) $\rightarrow$ (s $\vee \neg$ p)	2,3; rule of inference corresponding to rule 26a
5. p $\rightarrow$ (s $\vee \neg$ p)	1,4; hypothetical syllogism
6. $\neg$ p $\vee$ (s $\vee \neg$ p)	5; implication rule 10a
7. $\neg$ p $\vee$ ($\neg$ p $\vee$ s)	6; commutative law 2a
8. ($\neg$ p $\vee \neg$ p) $\vee$ s	7; associative law 3a
9. $\neg$ p $\vee$ s	8; idempotent law 5a
10. p $\rightarrow$ s	9; implication rule 10a

13. (a)

1. A $\wedge \neg$ B	hypothesis
2. A	1; rule 29
3. A $\rightarrow$ P	hypothesis
4. P	2,3; modus ponens
5. $\neg$ B	1; rule 29
6. P $\wedge \neg$ B	4,5; rule 34

(b)

1. $H \wedge \neg R$	hypothesis
2. H	1; rule 29
3. $\neg \neg N$	negation of conclusion
4. N	3; rule 1
5. $H \wedge N$	2,4; rule 34
6. $(H \wedge N) \to R$	hypothesis
7. R	5,6; modus ponens
8. $\neg R$	1; rule 29
9. $R \wedge \neg R$	7,8; rule 34
10. contradiction	9; rule 7b

14. The hypotheses are:

$H \to (T = F)$, $\neg C \to (T = S)$, $\neg (T = Y) \vee (T = F)$, $H \to \neg C$, $\neg (F = S)$,

$(T = S) \to (Y = F)$ and, implicitly, $(x = y) \to (y = x)$, $(x = y) \wedge (y = z) \to (x = z)$.

(a) No. $H \to \neg C \to (T = S)$ gives $H \to (T = F) \wedge (T = S) \to (F = S)$. But

$\neg (F = S)$, so [modus tollens] $\neg H$.

(b) Yes. Assume $\neg C$. Then $T = S$, so $Y = F$. Suppose that $Y = T$. Then $T = S$ and $Y = T$, so $Y = S$. Hence $F = Y = S$, a contradiction. Thus $\neg (Y = T)$.

(c) Friday, of course, as in (b).

CHAPTER 3

Relations are the unifying theme of this chapter, which also introduces graphs and digraphs, matrices and modular arithmetic. We want students to be able to think of relations geometrically as digraphs and algebraically as matrices.

Section 3.1 introduces the concept of relation as a set of ordered pairs, gives some examples, and suggests that pictures may be helpful. This is a good section to use for practice with definitions. What do these words really mean? What are some **non**examples of symmetric relations? Does every relation have to be either symmetric or antisymmetric? Etc.

Section 3.2 gives the basic definitions of graph theory. We take up digraphs first because of their link with relations in § 3.1. The reachable and adjacency relations will be important later on when we look at graphs as potential data structures. It's worth noting in class that one reason we study graphs is that they give us a conceptual geometric framework for modeling constructs, such as relations, that we have no inherent picture for. Computers don't draw graphs, but people do in order to visualize what the computers are dealing with.

Section 3.3 presents matrices as arrays. It's pretty standard stuff. Examples 4-7 tie matrices to relations and graphs. Matrices whose entries are given by formulas, such as the ones in Exercises 5 and 6, often cause students trouble. Assign one of these exercises and give some examples in class. It's not too late to kick out any students who think the singular of matrices is "matricie".

In § 3.4 we motivate the definition of matrix multiplication by counting paths in a digraph, and we then simply give the definition. Work out a couple of entries and exhibit the corresponding paths of length 2 in your digraph. Our own view is that the **right** way to see associativity of matrix multiplication is in the context of linear transformations, which we have no other reason to study here. So we simply say in a loud voice that multiplication is associative, and give examples. Exercises 5, 7 and 22 help convince students. Exercises 18-23 are fairly hard for many students. It's good to assign a few of them, though.

Section 3.5 gives the **whole story** on equivalence relations, which we view as another way of thinking about partitions. The characterization in Theorem 2 is fundamental. Make sure students really understand Example 10. It works well to take a few examples and view them from the three perspectives: relations between elements [still an intuitive idea at this point], partitions, and constant-value sets for functions.

Well-definedness of functions is a very slippery concept for many students. Since the question is always linked to some sort of equivalence relation, this is probably the first chance the students have had to really understand it, and this is the place to explain what's at issue.

Section 3.6 serves two purposes. Its main theme is modular arithmetic on the set $\mathbb{Z}(p)$ of congruence classes mod p, but it also gives a statement of the Division Algorithm [or, as you may prefer to call it, the Division Property] on $\mathbb{Z}$. The treatment of $\mathbb{Z}(p)$ gives an important example of an equivalence relation and presents the mechanics of the DIV and MOD functions that students encounter in computer science and that we'll use heavily in Chapter 12. Even if you choose to soft-pedal $\mathbb{Z}(p)$ in a one-semester course, be sure to mention the Division Algorithm, since it forms the subject matter for the introduction to loop invariants in § 4.1.

Section 3.1

1. (a) R_1 satisfies (AR) and (S).
   (b) R_2 satisfies (R), (S) and (T).
   (c) R_3 satisfies (R), (AS) and (T).
   (d) R_4 satisfies only (S).
   (e) R_5 satisfies only (S).

2. (a) $\{(0,0),\ (0,1),\ (0,2),\ (1,1),\ (1,2),\ (2,2)\}$.
   (b) $\{(0,1),\ (0,2),\ (1,2)\}$.
   (c) $\{(0,0),\ (1,1),\ (2,2)\}$.
   (d) $\{(0,0),\ (0,1),\ (0,2),\ (1,0),\ (2,0)\}$.
   (e) $\{(0,0),\ (0,1),\ (0,2),\ (1,1),\ (2,1)\}$.
   (f) $\{(0,0),\ (0,1),\ (0,2),\ (1,0),\ (1,1),\ (2,0)\}$.
   (g) $\{(1,1)\}$.
   (h) $\emptyset$.
   (i) $\{(1,0),\ (1,1),\ (2,2)\}$.

3. The relations in (a) and (c) are reflexive. The relations in (c), (d), (f), (g) and (h) are symmetric.

4. (a) $\{(0,5),\ (1,4),\ (2,3),\ (3,2),\ (4,1),\ (5,0)\}$.
   (b) $\{(0,2),\ (1,2),\ (2,2),\ (2,1),\ (2,0)\}$.

34

(c) $(2,2)$, $(73,2)$ and $(2,17)$ are some.

5. R_1 satisfies (AR) and (S). R_2 and R_3 satisfy only (S).

6. R satisfies (R), (S) and (T).

7. (a) The empty relation satisfies (AR), (S), (AS) and (T). The last three properties hold vacuously.
(b) U satisfies (R), (S) and (T).

8. (a) The relation $<$ on any set, say $\{1,2\}$.
(b) The relation $\neq$ on any set, say $\{1,2\}$.

9. (a) If $E \subseteq R_1$ and $E \subseteq R_2$, then $E \subseteq R_1 \cap R_2$. Alternatively, if R_1 and R_2 are reflexive and $x \in S$ then $(x,x) \in R_1$ and $(x,x) \in R_2$, so $(x,x) \in R_1 \cap R_2$.
(b) If R_1 and R_2 are symmetric and $(x,y) \in R_1 \cap R_2$, then $(x,y) \in R_1$ and $(x,y) \in R_2$, so $(y,x) \in R_1$ and $(y,x) \in R_2$.
(c) Suppose R_1 and R_2 are transitive. If $(x,y), (y,z) \in R_1 \cap R_2$ then $(x,y), (y,z) \in R_1$ and so $(x,z) \in R_1$. Similarly $(x,z) \in R_2$.

10. (a) Yes. This is clear: if $E \subseteq R_1$ and $E \subseteq R_2$ then $E \subseteq R_1 \cup R_2$.
(b) Yes. If $(x,y) \in R_1 \cup R_2$ then $(x,y) \in R_1$ or $(x,y) \in R_2$, and in either case $(y,x) \in R_1 \cup R_2$.
(c) No. For a small example, let $S = \{a, b, c\}$, $R_1 = \{(a,b)\}$ and $R_2 = \{(b,c)\}$.

11. (a) Suppose R is symmetric. If $(x,y) \in R$, then $(y,x) \in R$ by symmetry and so $(x,y) \in R^{\leftarrow}$. Similarly $(x,y) \in R^{\leftarrow}$ implies $(x,y) \in R$ [check] so that $R = R^{\leftarrow}$. For the converse, suppose that $R = R^{\leftarrow}$ and show R is symmetric.
(b) Suppose R is antisymmetric and $(x,y) \in R \cap R^{\leftarrow}$. Then $(x,y) \in R$ and $(y,x) \in R$, so $x = y$ and $(x,y) \in E$. Conversely, if $R \cap R^{\leftarrow} \subseteq E$ and if $(x,y) \in R$ and $(y,x) \in R$, then $(x,y) \in R \cap R^{\leftarrow} \subseteq E$ and so $x = y$.

12. (a) If $(t,s) \in (R_1 \cup R_2)^{\leftarrow}$ then $(s,t) \in R_1 \cup R_2$, so $(s,t) \in R_1$ or $(s,t) \in R_2$, whence $(t,s) \in R_1^{\leftarrow}$ or $(t,s) \in R_2^{\leftarrow}$ and thus $(t,s) \in R_1^{\leftarrow} \cup R_2^{\leftarrow}$. This argument is reversible, so $(R_1 \cup R_2)^{\leftarrow} \supseteq R_1^{\leftarrow} \cup R_2^{\leftarrow}$ as well.
(b) Replace "or" by "and" in the answer to part (a).

(c) Suppose $R_1 \subseteq R_2$ and $(t,s) \in R_1^{\leftarrow}$. Then $(s,t) \in R_1 \subseteq R_2$, so $(s,t) \in R_2$ and thus $(t,s) \in R_2^{\leftarrow}$.

13. (a)

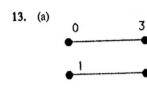

(b)

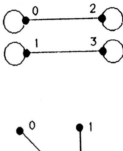

(c) See Figure 1(a).

(d)

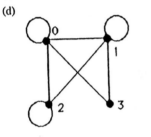

(e)

14. (a)

(b) Omit the loops from the answer to (a).

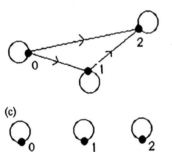

(c)

(d)

(e)

(f)

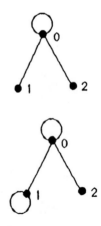

36

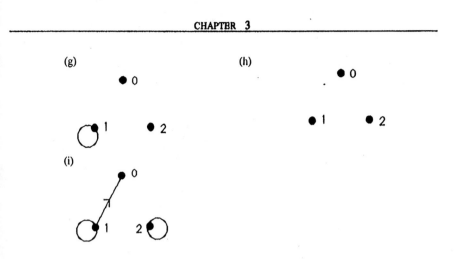

(g)

● 0

1 ● 2

(h)

● 0

● 1   ● 2

(i)

● 0

● 1   2 ●

Section 3.2

1. (a)

e	a	b	c	d	e	f
$\gamma(e)$	(x,v)	(v,x)	(v,w)	(w,y)	(w,y)	(y,x)

(b)

e	a	b	c	d	e	f	g	h
$\gamma(e)$	(u,v)	(u,x)	(v,w)	(v,y)	(x,w)	(x,y)	(w,z)	(y,z)

(c)

e	a	b	c	d
$\gamma(e)$	(x,w)	(w,x)	(y,z)	(z,y)

(d)

e	a	b	c	d	e
$\gamma(e)$	(x,x)	(y,y)	(z,z)	(x,y)	(y,z)

2.

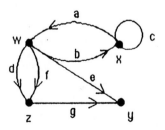

3. (a) Yes.   (b) Yes.   (c) No. There is no edge from t to x.
   (d) No. There is no edge from y to s.   (e) Yes.
   (f) No. There is no edge from u to x.

4. The shortest path is x z w, of length 2.

5. (a) x w y or x w v z y.   (b) y v z or y x w v z.
   (c) v x w or v z w or v z x w or v z y x w.
   (d) w v z.   (e) z y x w v or z w v or z x w v or z y v.

6.

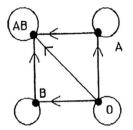

This digraph contains loops, so it is not acyclic.

7. Here is one:

8. (a) The reachability relation is the universal relation, i.e., all pairs of vertices.
   (c) R = {(w,w), (w,x), (x,x), (y,y), (y,z), (z,z)}.
   (d) R = {(x,x), (x,y), (x,z), (y,y), (y,z), (z,z)}.

9. (a) (v,w), (v,y), (v,z). Note that (v,z) is in the reachability relation but not in the adjacency relation.
   (b) (v,s), (v,t), (v,u), (v,w).
   (c) All of them. The reachability relation is the universal relation.

10. All describe paths except for parts (d) and (e). In (d), {z, z} is not an edge. In (e), {w, z} is not an edge.

11. (a) 2. (b) 7. (c) 3. (f) 2. (g) 3.
Parts (d) and (e) are not vertex sequences for paths.

12. Parts (b) and (f) describe closed paths. Remember, (e) does not describe a path.

13. (a) Edges e, f and g are parallel.
(b) Edges a and b are parallel. So are the loops e and f.

14. (a) 2. (b) 3.

15. (a) $A = \{(w,w), (w,x), (x,w), (y,y), (w,y), (y,w), (x,z), (z,x)\}$, while R consists of all sixteen ordered pairs of vertices.
(b) $A = \{(v,w), (w,v), (v,y), (y,v), (w,w), (z,z)\}$ and
$R = \{(z,z), (v,v), (v,w), (v,y), (w,v), (w,w), (w,y), (y,v), (y,w), (y,y)\}$.

16. (a) d e, d f or d g. Each has vertex sequence w x z.
(b) There are 90 possible answers. Two examples are e d d g, with vertex sequence z x w x z, and e e f e, with vertex sequence z x z x z.
(c) There are 9 possible answers, all with vertex sequence z x w w x z.
(d) There are 12 possible answers, such as b b d with vertex sequence w y w x.

17. (a) c a d or c b d. Both have vertex sequence y v w w.
(b) c c c with vertex sequence v y v y. There are 4 other such paths, each with vertex sequence v w v y.
(c) There are 4 such paths, each with vertex sequence v w w v y.
(d) There are 8 such paths, like e f f. They all have vertex sequence z z z z.

Section 3.3

1. (a) 1. (b) 5. (c) 2. (d) 0.

2. (a) 2. (b) 3. (c) 1. (d) 8.

3. (a) $\begin{bmatrix} -1 & 1 & 4 \\ 0 & 3 & 2 \\ 2 & -2 & 3 \end{bmatrix}$. (b) $\begin{bmatrix} 1 & 2 & 5 \\ 3 & -4 & -2 \end{bmatrix}$. (c) $\begin{bmatrix} 5 & 8 & 7 \\ 5 & 1 & 5 \\ 7 & 3 & 5 \end{bmatrix}$.
(d) Does not exist. (e) $\begin{bmatrix} 5 & 5 & 7 \\ 8 & 1 & 3 \\ 7 & 5 & 5 \end{bmatrix}$. (f) $\begin{bmatrix} 5 & 5 & 7 \\ 8 & 1 & 3 \\ 7 & 5 & 5 \end{bmatrix}$.

(g) $\begin{bmatrix} 12 & 12 & 8 \\ 12 & -4 & 8 \\ 8 & 8 & 4 \end{bmatrix}$. (h) Does not exist. (i) $\begin{bmatrix} 4 & 8 & 9 \\ 6 & 4 & 3 \\ 11 & 5 & 8 \end{bmatrix}$.

4. (a) $(1,-1,1)$. (b) $(2,0,-1)$. (c) $(1,-1,0)$. (d) $(2,-1,1)$.

5. (a) $\begin{bmatrix} 1 & -1 & 1 & -1 \\ -1 & 1 & -1 & 1 \\ 1 & -1 & 1 & -1 \end{bmatrix}$. (b) $\begin{bmatrix} 3 & 2 & 5 \\ 2 & 5 & 4 \\ 5 & 4 & 7 \\ 4 & 7 & 6 \end{bmatrix}$. (c) Not defined.

(d) $\begin{bmatrix} 3 & 2 & 5 & 4 \\ 2 & 5 & 4 & 7 \\ 5 & 4 & 7 & 6 \end{bmatrix}$. (e) $\begin{bmatrix} 3 & 2 & 5 & 4 \\ 2 & 5 & 4 & 7 \\ 5 & 4 & 7 & 6 \end{bmatrix}$. (f) $\begin{bmatrix} 2 & -2 & 2 \\ -2 & 2 & -2 \\ 2 & -2 & 2 \\ -2 & 2 & -2 \end{bmatrix}$.

6. (a) $\begin{bmatrix} 3 & 7 & 13 \\ 5 & 10 & 17 \\ 7 & 13 & 21 \end{bmatrix}$. (b) 14. (c) $9 + 18 + 33 = 17 + 20 + 23 = 60$.

(d) $B[1,2] + B[1,3] + B[2,2] + B[2,3] = 5 + 10 + 6 + 11 = 32$;
$B[2,1] + B[3,1] + B[2,2] + B[3,2] = 3 + 4 + 6 + 7 = 20$.

(e) $(A[1,2] + A[1,3]) \cdot (A[2,2] + A[2,3]) = (2 + 3) \cdot (4 + 6) = 50$.

(f) Yes. (g) No.

7. (a) $\begin{bmatrix} 1 & 0 & 0 \\ 0 & 1 & 0 \\ 0 & 0 & 1 \end{bmatrix}$, $\begin{bmatrix} 1 & 0 & 0 \\ 0 & 0 & 1 \\ 0 & 1 & 0 \end{bmatrix}$, $\begin{bmatrix} 0 & 1 & 0 \\ 1 & 0 & 0 \\ 0 & 0 & 1 \end{bmatrix}$, $\begin{bmatrix} 0 & 1 & 0 \\ 0 & 0 & 1 \\ 1 & 0 & 0 \end{bmatrix}$, $\begin{bmatrix} 0 & 0 & 1 \\ 1 & 0 & 0 \\ 0 & 1 & 0 \end{bmatrix}$, $\begin{bmatrix} 0 & 0 & 1 \\ 0 & 1 & 0 \\ 1 & 0 & 0 \end{bmatrix}$.

(b) Four of them equal their transposes.

8. All are true.

9. (a) $\begin{bmatrix} 1 & 0 \\ n & 1 \end{bmatrix}$. (b) $\{0\}$.

(c) $\{n \in \mathbb{N} : n \text{ is odd}\}$. (d) $\{n \in \mathbb{N} : n \text{ is even}\}$.

10. (a) $(A - B) + B = (A + (-B)) + B$ [definition of $A - B$]
$= A + ((-B) + B)$ [associative law]
$= A + 0$ [additive inverses]
$= A$ [additive identity].

Or give an argument as in the answer to part (b) below.

(b) $(-(A - B))[i,j] = -((A - B)[i,j]) = -((A + (-B))[i,j])$
$= -(A[i,j] + (-B)[i,j]) = -(A[i,j] - B[i,j]) = -A[i,j] + B[i,j]$
$= B[i,j] + (-A)[i,j] = (B - A)[i,j]$ for all i,j.

(c) Any example with $C \neq 0$ will work.

11. (a) The (i,j) entry of $a\mathbf{A}$ is $a\mathbf{A}[i,j]$. Similarly for $b\mathbf{B}$, and so the (i,j) entry of $a\mathbf{A} + b\mathbf{B}$ is $a\mathbf{A}[i,j] + b\mathbf{B}[i,j]$. So the (i,j) entry of $c(a\mathbf{A} + b\mathbf{B})$ is $ca\mathbf{A}[i,j] + cb\mathbf{B}[i,j]$. A similar discussion shows that this is the (i,j) entry of $(ca)\mathbf{A} + (cb)\mathbf{B}$. Since their entries are equal, the matrices $c(a\mathbf{A} + b\mathbf{B})$ and $(ca)\mathbf{A} + (cb)\mathbf{B}$ are equal.

(b) For all i, j we have $(-a\mathbf{A})[i,j] = -(a\mathbf{A})[i,j] = -(a\mathbf{A}[i,j]) = (-a)\mathbf{A}[i,j] = ((-a)\mathbf{A})[i,j]$ and also $-(a\mathbf{A}[i,j]) = a(-\mathbf{A}[i,j]) = a((-\mathbf{A})[i,j]) = (a(-\mathbf{A}))[i,j]$.

(c) The (j,i) entries of both $(a\mathbf{A})^{\mathrm{T}}$ and $a\mathbf{A}^{\mathrm{T}}$ equal $a\mathbf{A}[i,j]$. Here $1 \le i \le m$ and $1 \le j \le n$. So the matrices are equal.

12. (b) For all i, j $(\mathbf{A} + \mathbf{B})[i,j] = \mathbf{A}[i,j] + \mathbf{B}[i,j] = \mathbf{B}[i,j] + \mathbf{A}[i,j] = (\mathbf{B} + \mathbf{A})[i,j]$. (c) and (d) have similar proofs.

13. (a) $\begin{bmatrix} 0 & 0 & 1 & 0 \\ 0 & 0 & 1 & 0 \\ 0 & 0 & 0 & 0 \\ 0 & 0 & 1 & 0 \end{bmatrix}$. (b) $\begin{bmatrix} 1 & 1 & 0 & 0 \\ 1 & 1 & 0 & 0 \\ 0 & 0 & 0 & 2 \\ 0 & 0 & 0 & 0 \end{bmatrix}$. (c) $\begin{bmatrix} 0 & 0 & 0 & 0 \\ 0 & 0 & 0 & 2 \\ 0 & 0 & 1 & 0 \\ 0 & 1 & 0 & 0 \end{bmatrix}$.

14. (a) $\begin{bmatrix} 2 & 1 & 1 & 0 \\ 1 & 1 & 0 & 0 \\ 1 & 0 & 0 & 1 \\ 0 & 0 & 1 & 0 \end{bmatrix}$. (b) $\begin{bmatrix} 0 & 1 & 1 & 1 & 0 \\ 1 & 0 & 0 & 0 & 1 \\ 1 & 0 & 0 & 1 & 1 \\ 1 & 0 & 1 & 0 & 1 \\ 0 & 1 & 1 & 1 & 0 \end{bmatrix}$. (c) $\begin{bmatrix} 0 & 1 & 1 & 0 & 0 \\ 1 & 0 & 0 & 1 & 1 \\ 1 & 0 & 0 & 1 & 1 \\ 0 & 1 & 1 & 0 & 1 \\ 0 & 1 & 1 & 1 & 0 \end{bmatrix}$. (d) $\begin{bmatrix} 0 & 1 & 0 & 0 \\ 1 & 1 & 0 & 0 \\ 0 & 0 & 1 & 0 \\ 0 & 0 & 0 & 0 \end{bmatrix}$.

15. (a)   (b)   (c)

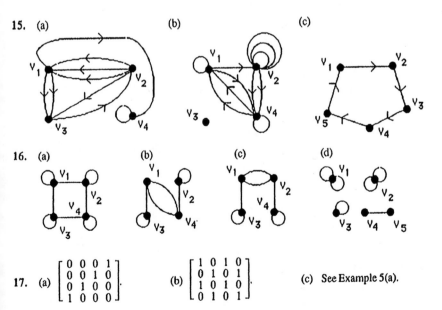

16. (a)   (b)   (c)   (d)

17. (a) $\begin{bmatrix} 0 & 0 & 0 & 1 \\ 0 & 0 & 1 & 0 \\ 0 & 1 & 0 & 0 \\ 1 & 0 & 0 & 0 \end{bmatrix}$. (b) $\begin{bmatrix} 1 & 0 & 1 & 0 \\ 0 & 1 & 0 & 1 \\ 1 & 0 & 1 & 0 \\ 0 & 1 & 0 & 1 \end{bmatrix}$. (c) See Example 5(a).

(d) $\begin{bmatrix} 1 & 1 & 1 & 1 \\ 1 & 1 & 1 & 1 \\ 1 & 1 & 1 & 0 \\ 1 & 1 & 0 & 0 \end{bmatrix}$.

(e) $\begin{bmatrix} 0 & 0 & 0 & 1 \\ 0 & 0 & 0 & 1 \\ 0 & 0 & 0 & 1 \\ 1 & 1 & 1 & 1 \end{bmatrix}$.

18.

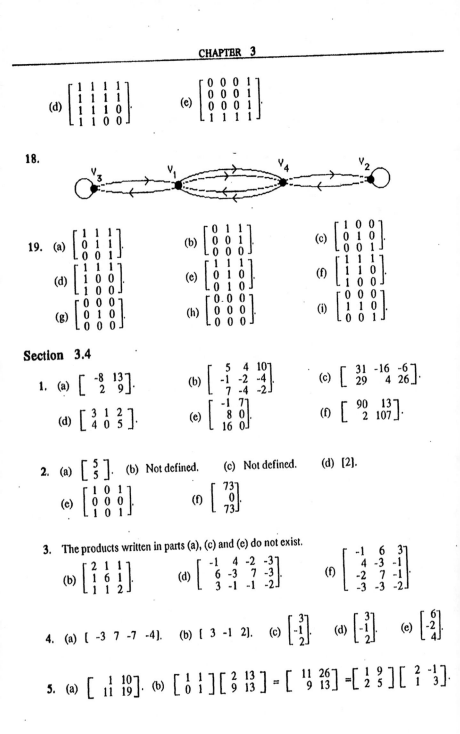

19. (a) $\begin{bmatrix} 1 & 1 & 1 \\ 0 & 1 & 1 \\ 0 & 0 & 1 \end{bmatrix}$.

(b) $\begin{bmatrix} 0 & 1 & 1 \\ 0 & 0 & 1 \\ 0 & 0 & 0 \end{bmatrix}$.

(c) $\begin{bmatrix} 1 & 0 & 0 \\ 0 & 1 & 0 \\ 0 & 0 & 1 \end{bmatrix}$.

(d) $\begin{bmatrix} 1 & 1 & 1 \\ 1 & 0 & 0 \\ 1 & 0 & 0 \end{bmatrix}$.

(e) $\begin{bmatrix} 1 & 1 & 1 \\ 0 & 1 & 0 \\ 0 & 1 & 0 \end{bmatrix}$.

(f) $\begin{bmatrix} 1 & 1 & 1 \\ 1 & 1 & 0 \\ 1 & 0 & 0 \end{bmatrix}$.

(g) $\begin{bmatrix} 0 & 0 & 0 \\ 0 & 1 & 0 \\ 0 & 0 & 0 \end{bmatrix}$.

(h) $\begin{bmatrix} 0 & 0 & 0 \\ 0 & 0 & 0 \\ 0 & 0 & 0 \end{bmatrix}$.

(i) $\begin{bmatrix} 0 & 0 & 0 \\ 1 & 1 & 0 \\ 0 & 0 & 1 \end{bmatrix}$.

Section 3.4

1. (a) $\begin{bmatrix} -8 & 13 \\ 2 & 9 \end{bmatrix}$.

(b) $\begin{bmatrix} 5 & 4 & 10 \\ -1 & -2 & -4 \\ 7 & -4 & -2 \end{bmatrix}$.

(c) $\begin{bmatrix} 31 & -16 & -6 \\ 29 & 4 & 26 \end{bmatrix}$.

(d) $\begin{bmatrix} 3 & 1 & 2 \\ 4 & 0 & 5 \end{bmatrix}$.

(e) $\begin{bmatrix} -1 & 7 \\ 8 & 0 \\ 16 & 0 \end{bmatrix}$.

(f) $\begin{bmatrix} 90 & 13 \\ 2 & 107 \end{bmatrix}$.

2. (a) $\begin{bmatrix} 5 \\ 5 \end{bmatrix}$. (b) Not defined. (c) Not defined. (d) [2].

(e) $\begin{bmatrix} 1 & 0 & 1 \\ 0 & 0 & 0 \\ 1 & 0 & 1 \end{bmatrix}$.

(f) $\begin{bmatrix} 73 \\ 0 \\ 73 \end{bmatrix}$.

3. The products written in parts (a), (c) and (e) do not exist.

(b) $\begin{bmatrix} 2 & 1 & 1 \\ 1 & 6 & 1 \\ 1 & 1 & 2 \end{bmatrix}$.

(d) $\begin{bmatrix} -1 & 4 & -2 & -3 \\ 6 & -3 & 7 & -3 \\ 3 & -1 & -1 & -2 \end{bmatrix}$.

(f) $\begin{bmatrix} -1 & 6 & 3 \\ 4 & -3 & -1 \\ -2 & 7 & -1 \\ -3 & -3 & -2 \end{bmatrix}$.

4. (a) [-3 7 -7 -4]. (b) [3 -1 2]. (c) $\begin{bmatrix} 3 \\ -1 \\ 2 \end{bmatrix}$. (d) $\begin{bmatrix} 3 \\ -1 \\ 2 \end{bmatrix}$. (e) $\begin{bmatrix} 6 \\ -2 \\ 4 \end{bmatrix}$.

5. (a) $\begin{bmatrix} 1 & 10 \\ 11 & 19 \end{bmatrix}$. (b) $\begin{bmatrix} 1 & 1 \\ 0 & 1 \end{bmatrix}\begin{bmatrix} 2 & 13 \\ 9 & 13 \end{bmatrix} = \begin{bmatrix} 11 & 26 \\ 9 & 13 \end{bmatrix} = \begin{bmatrix} 1 & 9 \\ 2 & 5 \end{bmatrix}\begin{bmatrix} 2 & -1 \\ 1 & 3 \end{bmatrix}$.

6. (a) $AB = \begin{bmatrix} -1 & 3 \\ 2 & 7 \end{bmatrix}$ $BA = \begin{bmatrix} 1 & 9 \\ 2 & 5 \end{bmatrix}$.

   (b) $AC = \begin{bmatrix} 2 & 13 \\ 9 & 13 \end{bmatrix}$ $CA = \begin{bmatrix} -4 & 3 \\ 5 & 19 \end{bmatrix}$.   (c) $\begin{bmatrix} 9 & 16 \\ 8 & 33 \end{bmatrix}$.

7. (a) $\begin{bmatrix} 7 & 14 \\ 8 & 11 \\ 2 & -6 \end{bmatrix}$.   (b) $\begin{bmatrix} 3 & -1 \\ 2 & 1 \\ -2 & 4 \end{bmatrix}\begin{bmatrix} 1 & 4 \\ 0 & 1 \end{bmatrix} = \begin{bmatrix} 3 & 11 \\ 2 & 9 \\ -2 & -4 \end{bmatrix} = \begin{bmatrix} 3 & 5 \\ 2 & 5 \\ -2 & 0 \end{bmatrix}\begin{bmatrix} 1 & 2 \\ 0 & 1 \end{bmatrix}$.

9. (a) 2.   (b) 1.   (c) 2.   (d) 0.

10. (a) a a and g f.   (b) a g.   (c) a h and g k.   (d) There are none.

11. (a) $M^3 = \begin{bmatrix} 3 & 20 & 2 & 3 \\ 0 & 8 & 0 & 0 \\ 2 & 9 & 1 & 2 \\ 0 & 4 & 0 & 0 \end{bmatrix}$.   (b) 9.

   (c) f a b, f a c, f b d, f b e, f c d, f c e, f h j, k j d, k j e.

12. (a) Simply remove the arrows from Figure 1.
   (b) M^2 is given in Example 3. So there are $M^2[3,3] = 5$ paths of length 2 from v_3 to itself.
   (c) f f, g g, f g, g f, k k.
   (d) M^3 is given in Example 3, so answer is $M^3[3,3] = 8$.
   (e) f a f, g a g, f a g, g a f, f h k, g h k, k h f, k h g.

13. (a) Simply remove the arrows from Figure 1.
   (b) M^2 is given in Example 3. So the answer is $M^2[2,2] = 9$.
   (c) d d, e e, d e, e d, b b, c c, b c, c b, j j.
   (d) M^3 is given in Example 3, so the answer is $M^3[2,2] = 36$.

14. If $ad - bc \neq 0$, then check that the suggested matrix A^{-1} is the inverse of A. Conversely, suppose that A has an inverse; say

$$\begin{bmatrix} a & b \\ c & d \end{bmatrix}\begin{bmatrix} x & y \\ z & w \end{bmatrix} = \begin{bmatrix} 1 & 0 \\ 0 & 1 \end{bmatrix}.$$

Then $ax + bz = 1$ so a and b are not both 0. Moreover, we have $ay + bw = 0$ and $cy + dw = 1$, so $acy + bcw = 0$, $acy + adw = a$, and thus $(ad - bc)w = a$. Similarly, $ady + bdw = 0$ and $bcy + bdw = b$, so $(ad - bc)y = -b$. Since a and $-b$ are not both 0, $ad - bc \neq 0$.

15. (a) $I^{-1} = I$.   (b) $A^{-1} = \begin{bmatrix} 1 & -1 \\ 0 & 1 \end{bmatrix}$ (c) Not invertible.

    (d) $C^{-1} = \frac{1}{31} \begin{bmatrix} 8 & 3 \\ -5 & 2 \end{bmatrix}$ (e) $D^{-1} = D$.

16. Since $(A + B)(A - B) = A^2 + BA - AB - B^2$, it is enough to find A and B where $BA \neq AB$.

17. (b) Correct guess $A^n = \begin{bmatrix} 1 & 0 \\ n & 1 \end{bmatrix}$. Observe that

$$A^n A = \begin{bmatrix} 1 & 0 \\ n & 1 \end{bmatrix} \begin{bmatrix} 1 & 0 \\ 1 & 1 \end{bmatrix} = \begin{bmatrix} 1 & 0 \\ n+1 & 1 \end{bmatrix} = A^{n+1}.$$

18. Here we need $m = n$. For all i, k, $((aA)B)[i,k] = \sum_{j=1}^{n} (aA)[i,j]B[j,k] =$

$\sum_{j=1}^{n} aA[i,j]B[j,k] = a \sum_{j=1}^{n} A[i,j]B[j,k] = a(AB)[i,k] = (a(AB))[i,k]$ and also

$a \sum_{j=1}^{n} A[i,j]B[j,k] = \sum_{j=1}^{n} A[i,j]aB[j,k] = \sum_{j=1}^{n} A[i,j](aB)[j,k] = (A(aB))[i,k]$.

19. For $1 \leq k \leq p$ and $1 \leq i \leq m$,

$$(B^T A^T)[k,i] = \sum_{j=1}^{n} B^T[k,j]A^T[j,i] = \sum_{j=1}^{n} B[j,k]A[i,j].$$

Compare with the (k,i)-entry of $(AB)^T$.

20. (a) If $A + C = B + C$, then $A[i,j] + C[i,j] = B[i,j] + C[i,j]$ for all i,j. So $A[i,j] = B[i,j]$ for all i,j by the cancellation law for real numbers. Thus $A = B$.

    (b) For example, $\begin{bmatrix} 1 & 1 \\ 1 & 1 \end{bmatrix} \begin{bmatrix} 1 & -1 \\ -1 & 1 \end{bmatrix} = \begin{bmatrix} 0 & 0 \\ 0 & 0 \end{bmatrix} = \begin{bmatrix} 0 & 0 \\ 0 & 0 \end{bmatrix} \begin{bmatrix} 1 & -1 \\ -1 & 1 \end{bmatrix}$ but

$\begin{bmatrix} 1 & 1 \\ 1 & 1 \end{bmatrix} \neq \begin{bmatrix} 0 & 0 \\ 0 & 0 \end{bmatrix}$.

21. (a) In fact, $AB = BA = aB$ for all B in $\mathfrak{M}_{2,2}$.

    (b) $AB = BA$ with $B = \begin{bmatrix} 1 & 0 \\ 0 & 0 \end{bmatrix}$ forces $\begin{bmatrix} a & 0 \\ c & 0 \end{bmatrix} = \begin{bmatrix} a & b \\ 0 & 0 \end{bmatrix}$ and so $b = c = 0$. So

$A = \begin{bmatrix} a & 0 \\ 0 & d \end{bmatrix}$. Now try $B = \begin{bmatrix} 0 & 1 \\ 0 & 0 \end{bmatrix}$.

22. (a) $\begin{bmatrix} a_1 & a_2 \\ a_3 & a_4 \end{bmatrix} \left\{ \begin{bmatrix} b_1 & b_2 \\ b_3 & b_4 \end{bmatrix} \begin{bmatrix} c_1 & c_2 \\ c_3 & c_4 \end{bmatrix} \right\}$

$$= \begin{bmatrix} a_1b_1c_1 + a_1b_2c_3 + a_2b_3c_1 + a_2b_4c_3 & a_1b_1c_2 + a_1b_2c_4 + a_2b_3c_2 + a_2b_4c_4 \\ a_3b_1c_1 + a_3b_2c_3 + a_4b_3c_1 + a_4b_4c_3 & a_3b_1c_2 + a_3b_2c_4 + a_4b_3c_2 + a_4b_4c_4 \end{bmatrix}$$

$$= \left\{ \begin{bmatrix} a_1 & a_2 \\ a_3 & a_4 \end{bmatrix} \begin{bmatrix} b_1 & b_2 \\ b_3 & b_4 \end{bmatrix} \right\} \begin{bmatrix} c_1 & c_2 \\ c_3 & c_4 \end{bmatrix}$$

(b) Are you kidding? Actually, it's not so bad if you use summation notation.
$(A(BC))[i,\ell] = \Sigma_j A[i,j](BC)[j,\ell] = \Sigma_j A[i,j] \Sigma_k B[j,k]C[k,\ell] = \Sigma_k \Sigma_j A[i,j]B[j,k]C[k,\ell]$
$= \Sigma_k (AB)[i,k]C[k,\ell] = ((AB)C)[i,\ell]$, but such manipulations with double sums haven't been discussed in this book.

23. (a) Consider $1 \leq i \leq m$ and $1 \leq k \leq p$ and compare the (i,k) entries of $(A + B)C$ and $AC + BC$.

(b) If A is $m \times n$, then B and C must both be $n \times r$ for the same r. For $1 \leq i \leq m$ and $1 \leq k \leq r$

$(A(B + C))[i,k] = \sum_{j=1}^{n} A[i,j](B + C)[j,k] = \sum_{j=1}^{n} A[i,j](B[j,k] + C[j,k]) =$

$\sum_{j=1}^{n} A[i,j]B[j,k] + \sum_{j=1}^{n} A[i,j]C[j,k] = (AB)[i,k] + (AC)[i,k] = (AB + AC)[i,k]$.

Section 3.5

1. (a) is an equivalence relation.   (b) $\perp$ is not reflexive or transitive.

(c) is not reflexive because some Americans do not live in any state.

(d) $\approx$ is not transitive.

(e) is not an equivalence relation because $\approx$ is not transitive.

(f) $\cong$ is an equivalence relation.

2. (a) $[L]$ is the family of all lines parallel to L, including L itself.

(c) For example, $[K.A. Ross] = [C.R.B. Wright] =$ the set of all residents of Oregon. The state of Oregon itself is not an equivalence class.

(f) $[p]$ is the set of all people that have the same mother as p does.

(b), (d) and (e) do not give equivalence relations.

3. Very much so.

4. (a) The relation $\equiv$ is an equivalence relation. Since $m - m = 0$ is even, $m \equiv m$. If $m - n$ is even, so is its negative $n - m$. If $m - n$ and $n - p$ are even, so is $(m - n) + (n - p) = m - p$.

(b) The relation $\approx$ is reflexive and symmetric, but not transitive. For example, $1 \approx 2$ and $2 \approx 3$ but $1 \not\approx 3$.

5. (a) The possibilities are

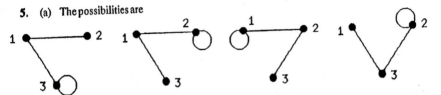

(b) For (R) no relabeling is required; $f = 1_{\{1, \ldots, n\}}$ works. For (S), if G can be labeled with f to become H, then reversing the labeling with f^{-1} turns H back into G. For (T), if f labels G to be H and g labels H to be K, then $g \circ f$ labels G to be K.

7. (a) Verify directly, or apply Theorem 2(a) with $f(m) = m^2$ for $m \in \mathbb{Z}$.

(b) There are infinitely many classes: $\{0\}$ and $\{n, -n\}$ for $n \in \mathbb{P}$.

8. Since $m - m = 0$, which is not odd, $m \sim m$ fails. If $m - n$ and $n - p$ are both odd, then $m - p = (m - n) + (n - p)$ is not odd.

9. (a) There are infinitely many classes: $\{0\}$ and the classes $\{n, -n\}$ for $n \in \mathbb{P}$.

(b) There are two classes:
$$h^{\leftarrow}(2) = \{n \in \mathbb{Z} : n \text{ is even}\} \text{ and } h^{\leftarrow}(0) = \{n \in \mathbb{Z} : n \text{ is odd}\}.$$

10. (a) Check (R), (S) and (T) or apply Theorem 2(a) with $f(m,n) = m - n$.

(b) $(m,n) \sim (k, \ell)$ if and only if $m - n = k - \ell$.

11. Apply Theorem 2, using the length function. The equivalence classes are the sets Σ^k, $k \in \mathbb{N}$.

12. Yes. One easily verifies (R), (S) and (T).

13. (a) Use brute force or Theorem 2(a) with part (b).

(b) Since $0 \notin \mathbb{P}$, $mq = np$ if and only if $m/n = p/q$.

14. No. Even if ν is one-to-one, so that ν^{-1} exists, ν^{-1} maps $[S]$ into S, not into $[S]$. The notation $\nu^{\leftarrow}([s])$ stands for $\{x \in S : \nu(x) = [s]\}$. More generally, if $\theta: A \to B$ is a function, then according to our definition preceding Example 6 of § 1.4, $\theta^{\leftarrow}(b) = \{a \in A : \theta(a) = b\}$ for each b in B. We can define a function, $\hat{\theta}$ say, from B to the set $\mathscr{P}(A)$ [but <u>not</u> to A itself] by $\hat{\theta}(b) = \theta^{\leftarrow}(b)$. In the present context, the equation $\nu^{\leftarrow}([s])$

= [s] translates into $\hat{\nu}([s]) = [s]$ with $\hat{\nu}: [S] \longrightarrow \mathscr{P}(S)$. Since $[S] \subseteq \mathscr{P}(S)$, $\hat{\nu}$ acts like the identity function on $[S]$. But $\hat{\nu}$ is not the same thing as ν^{-1}.

15. (a) Not well-defined: depends on the representative. For example $[3] = [-3]$ and $-3 \le 2$. If the definition made sense, we would have $[3] = [-3] \le [2]$ and hence $3 \le 2$.

(b) Trouble. For example, $[2] = [-2]$ but $(2)^2 + (2) + 1 \ne (-2)^2 + (-2) + 1$.

(c) Nothing wrong. If $[m] = [n]$ then $m^4 + m^2 + 1 = n^4 + n^2 + 1$.

(d) Not well-defined. For example, $[-1] = [1]$, but $[-1 + 1] = [0] \ne [2] = [1 + 1]$.

16. (a) Well-defined, since $f(x) = 1/x$ for $x \in \mathbb{Q}^+$.

(b) Not well-defined. For instance $1/2 = 3/6$, but $1^2 + 2^2 \ne 3^2 + 6^2$.

(c) Well-defined. Indeed, $h(m/n) = (m/n) + (n/m)$, so that $h(x) = x + 1/x$ for $x \in \mathbb{Q}^+$.

17. (a) Since $\sim$ is reflexive and symmetric, so is $\approx$. Given chains $f = f_1 \sim f_2 \sim \cdots \sim f_n = g$ and $g = g_1 \sim g_2 \sim \cdots \sim g_m = h$, we see that $f = f_1 \sim f_2 \sim \cdots \sim f_n \sim g_2 \sim \cdots \sim g_m = h$ is a chain, so $f \approx h$. Thus $\approx$ is transitive.

(b) The $\approx$-class of z is the set of all functions whose values are bounded above and below. If $z = f_1 \sim f_2 \sim \cdots \sim f_n$ then $|f_n(x)| \le n$ for every x in $[0,1]$.

18. One needs to show:

(R) For each (s_n), the set $\{n \in \mathbb{N} : s_n \ne s_n\}$ is finite, i.e., $(s_n) \sim (s_n)$.

(S) If $\{n \in \mathbb{N} : s_n \ne t_n\}$ is finite, so is $\{n \in \mathbb{N} : t_n \ne s_n\}$, i.e., $(s_n) \sim (t_n)$ implies $(t_n) \sim (s_n)$.

(T) If $\{n \in \mathbb{N} : s_n \ne t_n\}$ and $\{n \in \mathbb{N} : t_n \ne u_n\}$ are finite, so is $\{n \in \mathbb{N} : s_n \ne u_n\}$, i.e., $(s_n) \sim (t_n)$ and $(t_n) \sim (u_n)$ imply $(s_n) \sim (u_n)$.

Section 3.6

1. (a) $q = 6$, $r = 2$.   (b) $q = 5$, $r = 0$.   (c) $q = -7$, $r = 1$.

(d) $q = -5$, $r = 0$.   (e) $q = 5711$, $r = 31$.   (f) $q = -5712$, $r = 34$.

2. The answers are the same as for Exercise 1, with m DIV n instead of q, and m MOD n instead of r.

3. (a) $-4, 0, 4$. (b) $-3, 1, 5$. (c) $-2, 2, 6$. (d) $-1, 3, 7$. (e) $-4, 0, 4$.

4. (a) $[0]_4, [1]_4, [2]_4, [3]_4$.   (b) 73.

5. (a) 1.   (b) 3.   (c) 1.   (d) 2.   (e) 0.

6. (a) 1.   (b) 4.   (c) 2.   (d) k.   (e) k.

7. (a) 3 and 2.   (b) $m +_{10} k$ is the last [decimal] digit of $m + k$.
    (c) $m *_{10} k$ is the last [decimal] digit of $m * k$.

8. (a) $A_0 = \{-9, -6, -3, 0, 3, 6, 9\}$, $A_1 = \{-8, -5, -2, 1, 4, 7, 10\}$ and
    $A_2 = \{-10, -7, -4, -1, 2, 5, 8\}$.
    (b) $A_3 = A_0$, $A_4 = A_1$, $A_{73} = A_1$.

9.

$+_4$	0	1	2	3
0	0	1	2	3
1	1	2	3	0
2	2	3	0	1
3	3	0	1	2

$*_4$	0	1	2	3
0	0	0	0	0
1	0	1	2	3
2	0	2	0	2
3	0	3	2	1

10. Solutions are 5, 4, 3, 2 and 1, respectively.

11. Solutions are 1, 3, 2 and 4, respectively.

12. (a) (R) and (S) are clear. If $m \sim n$ and $n \sim p$, then $m^2 - p^2 = (m^2 - n^2) + (n^2 - p^2)$ is a
multiple of 3, so $m \sim p$. Thus (T) holds.
    (b) 0, 3, 6 and 72 are some.   (c) 1, 2 and 73 are some.
    (d) These are all there are, since $(3k)^2 - 0^2$ and $(3k \pm 1)^2 - 1^2$ are multiples of 3 and
every number in $\mathbb{N}$ can be written in the form 3k, 3k - 1 or 3k + 1.

13. (a) $m \equiv n \pmod 1$ for all $m, n \in \mathbb{Z}$. There is only one equivalence class.
    (b) The conditions $m = (m \text{ DIV } 1) \cdot 1 + m \text{ MOD } 1$ and $0 \le m \text{ MOD } 1 < 1$ force
$m \text{ MOD } 1 = 0$ and $m \text{ DIV } 1 = m$.
    (c) $0 = 0 +_1 0$ and $0 = 0 *_1 0$.

14. (a) Suppose that $m \equiv n \pmod p$. Then $m - n$ is a multiple of p. Hence
$(m - n)(m + n)$ is also a multiple of p. Thus $m^2 - n^2$ is a multiple of p, and so
$m^2 \equiv n^2 \pmod p$. Or apply Theorem 2 to $m \equiv n \pmod p$ and $m \equiv n \pmod p$.
    (b) f is well-defined, by (a).
    (c) g is well-defined. If $m \equiv n \pmod 6$, so that $m = n + 6k$ with $k \in \mathbb{Z}$, then
$m^2 = n^2 + 12nk + 36k^2 \equiv n^2 \pmod{12}$.

(d) h is not well-defined. For instance, $[1]_6 = [7]_6$ but $[1^3]_{12} = [1]_{12} \neq [7]_{12} = [7^3]_{12}$.

15. (a) $n = 1000a + 100b + 10c + d = a + b + c + d + 9 \cdot (111a + 11b + c) \equiv a + b + c + d$ (mod 9), or use Theorem 2 together with $1000 \equiv 100 \equiv 10 \equiv 1$ (mod 9).

(b) Sure. $10^k \equiv 1^k \equiv 1$ (mod 9) for every $k \in \mathbb{P}$, so $\sum_{k=0}^{m} a_k 10^k \equiv \sum_{k=0}^{m} a_k$ (mod 9) by Theorem 2.

16. (a) $n = 1000a + 100b + 10c + d \equiv d$ (mod 2) since $1000 \equiv 100 \equiv 10 \equiv 0$ (mod 2).

(b) Like (a).

17. Like Exercise 15. Note that $1000 = 91 \cdot 11 - 1$, $100 = 9 \cdot 11 + 1$, $10 = 1 \cdot 11 - 1$, so $1000a + 100b + 10c + d \equiv -a + b - c + d \equiv 0$ (mod 11) if and only if $a - b + c - d \equiv 0$ (mod 11).

18. (a) If n MOD $p = 0$ then $n = (n$ DIV $p) \cdot p$, so $-n = -(n$ DIV $p) \cdot p + 0$ and thus $(-n)$ DIV $p = -(n$ DIV $p)$ and $(-n)$ MOD $p = 0$.

(b) If $0 < n$ MOD $p < p$, then $0 < p - n$ MOD $p < p$ and $-n = -[(n$ DIV $p) \cdot p + n$ MOD $p] = [-(n$ DIV $p) - 1] \cdot p + [p - n$ MOD $p]$.

19. We have $q \cdot p - q' \cdot p = r' - r$, so $r' - r$ is a multiple of p. But $-p < -r \leq r' - r \leq r' < p$, and 0 is the only multiple of p between $-p$ and p. Thus $r' = r$, so $0 = (q - q') \cdot p$ and $q = q'$.

20. (a) If $\theta(m) = \theta(n)$ with $m, n \in \mathbb{Z}(p)$ then $[m]_p = [n]_p$, so $m \equiv n$ (mod p) and $m = m$ MOD $p = n$ MOD $p = n$. Thus θ is one-to-one. Since $[m]_p = [m$ MOD $p]_p = \theta(m$ MOD $p)$ and m MOD $p \in \mathbb{Z}(p)$, θ maps $\mathbb{Z}(p)$ onto $[\mathbb{Z}]_p$.

(b) Using Example 6, $\theta(m) + \theta(n) = [m]_p + [n]_p = [m + n]_p = [(m + n)$ MOD $p]_p = [m +_p n]_p = \theta(m +_p n)$. The proof for $\cdot$ is similar.

21. (a) By Theorem 3(a)
$$(m \text{ MOD } p) +_p (n \text{ MOD } p) = (m + n) \text{ MOD } p = (n + m) \text{ MOD } p$$
$$= (n \text{ MOD } p) +_p (m \text{ MOD } p).$$

Since $m, n \in \mathbb{Z}(p)$, m MOD $p = m$ and n MOD $p = n$.

(b) As in part (a), $(m +_p n) +_p r = (m \text{ MOD } p +_p n \text{ MOD } p) +_p r \text{ MOD } p$
$$= (m + n) \text{ MOD } p +_p r \text{ MOD } p = [(m + n) + r] \text{ MOD } p$$
and similarly $m +_p (n +_p r) = [m + (n + r)] \text{ MOD } p$. But $m + (n + r) = (m + n) + r$.

CHAPTER 4

Our computer science colleagues tell us that from their point of view induction is the most important technique in the discrete math course. Experience shows that even some of the better students still have trouble with induction at the end of a year's course, especially if the instructor has not really insisted that they learn it. Be firm.

We have completely overhauled our presentation of induction and recursion in this edition. Since our treatment is not the old familiar one, you will probably want to read through the whole chapter before starting to teach any of it. The core facts about induction are in §§ 4.2 and 4.5, with a recursive interlude between the two doses. The main innovations are our presentation of loop invariants and our careful account of recursion in its various aspects, which we begin in this chapter and continue in Chapter 7. We will be especially interested to hear how this chapter works with your students.

Section 4.1 introduces algorithm segments, called **while loops**, that occur in iterative algorithms. The main example of the section develops an algorithm for finding quotient and remainder in $\mathbb{N}$. The algorithm terminates because of the Well-Ordering Property, which we simply state as a fact about $\mathbb{N}$. The algorithm gives the right answer because "$q \cdot n + r = m$ and $r \geq 0$" is a loop invariant. We've learned from our students that computer science instructors use loop invariants a lot, often without explanation, so we've given both a careful definition and a clear statement of the fundamental theorem. The idea is completely natural and makes some of our algorithms easy to verify later on, too.

Section 4.2 introduces mathematical induction. One reason we started the chapter with loop invariants is so that we could use them to justify induction. If you want to give a more traditional justification based on the Well-Ordering Property you can do that instead, of course. We think it's wise to give some sort of mathematical justification just so students won't see induction as a trick out of the blue to use when you get stuck. An extra day spent on this section doing lots of examples [not just sums] might be a good idea.

Students need to understand that induction gives a framework or template for a proof, but it doesn't provide the details. Here is one way to present induction that seems to help many students. Think of a business letter format [perhaps produced by word processing software] with its date, address, salutation, body, closing and signature. Every letter has these parts, but what goes in the parts varies from letter to letter. Now draw the following on the blackboard:

p(m) is true
because · · ·

Let p(n) =

Assume p(k) is true

Here is an argument that p(k + 1) must
be true, assuming that p(k) is true

Thus by induction,
p(n) is true for every n ≥ m

Induction works like this. It's a framework for a valid mathematical proof, and it produces a correct proof if each essential part of it is correctly written.

Section 4.3 introduces recursion in the context of definition and calculation of sequences. The Fibonacci sequence makes a good example for class discussion. We return to it in the next section. Observe the difference between Exercises 11 and 14. The students can use induction as presented in § 4.2 to give a proof in Exercise 11(b), but 14(b) is considerably harder right now. [See the treatment in Example 2 of § 4.5.] We give a much more complete answer to 14(b) than we expect from students. Exercise 22 is pretty hard.

Section 4.4 just gives a taste of solving recurrence relations. You may want to spend a moment in class discussing how the theorem for second order linear recurrences might extend to higher order recurrences with constant coefficients. To go much farther in solving recurrence relations takes some fairly fancy machinery, of course, and our students may not have seen calculus, let alone differential equations or complex power series. The account of divide-and-conquer recurrences is included because they occur so commonly. Don't prove Theorem 2 in class; just illustrate its use for some monotone sequence. Perhaps ask a student to bring an example from a computer science class.

Present the Second Principle of Induction in § 4.5 as an improvement, as simple as the First Principle and more useful. Show how to do one of Exercises 6 - 9, before you start to prove inequalities.

Section 4.6 on the Euclidean Algorithm presents material that we won't use again until the end of the book. It's nice mathematics, though, and gives a beautiful illustration of how a loop invariant helps shape an algorithm. Any math majors in the class will surely want to go through this section, even if you omit it from the syllabus. The discussion of congruences ties this material back to our account of $\mathbb{Z}(p)$ in § 3.6.

Section 4.1

1. (a) 0, 3, 9, 21, 45. (b) 1, 5, 13, 29, 61. (c) 1, 1, 1, 1, 1.

2. (a) $b = 7$. (b) $b = 6$.

3. (a)

	m	n
initially	0	0
after first pass	1	1
after second pass	4	2
after third pass	9	3
after fourth pass	16	4

  (b) Replace 4 by 17.

4. (a) 0, 1, 4, 9, 16. (b) 16.

5. (a) 4, 16, 36, 64. (b) 9, 25, 49, 81, 121, ...

6. (a) i := 1
    while i < 18 do
        k := k + 2i
        i := i + 1

(b) k := 8
    while k > 0 do
        i := i + 2k
        k := k - 1

7. (a) If $m + n$ is even, so is $(m + 1) + (n + 1) = (m + n) + 2$. Of course, we didn't need the guard $1 \leq m$ to see this.
   (b) If $m + n$ is odd, so is $(m + 1) + (n + 1)$.

8. (a) If $n^2 \geq m^3$ and $m \geq 1$, then $(3n)^2 = 9n^2 \geq 9m^3 > 8m^3 = (2m)^3$.
   (b) If $2m^6 < n^4$, then $2(2m)^6 = 64 \cdot 2m^6 < 64n^4 < 81n^4 = (3n)^4$.

9. (a) Yes. If $i < j^2$ and $j \geq 1$ then $i + 2 < j^2 + 2 < j^2 + 2j + 1 = (j + 1)^2$.

(b) Yes. The case $i < j^2$ and $j = 0$ cannot happen, and otherwise the answer for (a) applies.

(c) No. Consider the case $i = j = 0$.

(d) No. Consider the case $i = j = 1$, for example.

10. (a) Yes. Suppose $k \geq 1$ and $k^2 \equiv 1 \pmod 3$. Since $(2k)^2 = 3k^2 + k^2$ and since $3k^2 \equiv 0 \pmod 3$ and $k^2 \equiv 1 \pmod 3$, $(2k)^2 \equiv 1 \pmod 3$ by Theorem 2, § 3.6.

(b) No. If $k = 1$, $k^2 = 1 \equiv 1 \pmod 4$ but $(2k)^2 = 4 \not\equiv 1 \pmod 4$.

11. (a) $b \geq 2$. (b) $b \in \{2, 3, 5\}$. (c) $b \in \mathbb{N}$.

12. Change the guard to $m \leq n$. The initialization can remain the same or change to $m := 2$. The new loop invariant is FACT $= (m - 1)!$.

13. No. The sequence "$k := k^2$, print k" changes the value of k. Algorithm A prints 1, 4 and stops. Algorithm B prints 1, 4, 9, 16, because "for" resets k each time.

14. No. Algorithm C prints 2, 4, 8. Algorithm D prints 2, 4.

15. (a) Yes. new $r < 73$ by definition of MOD.

(b) No. If $r = 5 \equiv 0 \pmod 5$, for instance, then new $r = 9 \not\equiv 0 \pmod 5$.

(c) This is an invariant vacuously, because $r \leq 0$ and $r > 0$ cannot both hold at the start of the loop.

16. (a) new $S = S + 2I + 1 = I^2 + 2I + 1 = (I + 1)^2 = (\text{new } I)^2$.

(b) new $S = S + 2I + 1 = I^2 + 1 + 2I + 1 = (I + 1)^2 + 1 = (\text{new } I)^2 + 1$.

(c) 73^2 since the Loop Invariant Theorem applies to $S = I^2$.

(d) $73^2 + 1$ since the Loop Invariant Theorem applies to $S = I^2 + 1$.

17. The sets in (a), (c) and (f) have smallest elements, (f) because $n! > 80^n$ for all large enough n by Example 3(b) of § 1.6.

The sets in (b) and (e) fail, because they aren't subsets of $\mathbb{N}$.

The set in (d) fails because it is empty.

18. (a) We would get 1 in each case. No.

(b) Same as (a).

19. (a) This is an invariant: if $r > 0$ and a, b and r are multiples of 5, then the new values b, r and b MOD r are multiples of 5. Note that b MOD $r = b - (b$ DIV $r) \cdot r$.

(b) This is not an invariant. For example, if $r = 1, a = 5$ and $b = 3$ then $a = 3$ after execution of the loop.

(c) This is an invariant: If $r < b$ and $r > 0$ on entry into the loop then new $r <$ new b, i.e., b MOD $r < r$, by definition of MOD.

(d) This is an invariant vacuously, because $r \leq 0$ and $r > 0$ cannot both hold at the start of the loop.

20. (a) Yes. If $4 \leq k$ and $5^k < k!$ then $5^{k+1} = 5 \cdot 5^k < 5 \cdot k! \leq (k + 1) \cdot k! = (k + 1)!$.

(b) No. For example, this fails for $k = 4$.

21. (a) If m is even $2^{(new\ k)} \cdot (new\ m) = 2^{k+1} \cdot \frac{m}{2} = 2^k \cdot m = n$.

(b) If m is even, the loop replaces m by $m/2$ and tries again. The chain $n > n/2 > n/4 > \cdots$ is a decreasing chain of positive numbers, so its members cannot all be integers. During some pass $m/2$ cannot be an integer; i.e., m is odd. Then the algorithm exits the loop.

22. (a) Yes. If $p \wedge q$ holds at the start of the loop, then p is true and q is true. If g is also true, then since p and q are invariants of the loop they are both true at the end; hence $p \wedge q$ is also true at the end.

(b) Yes. If $p \vee q$ holds at the start then, say, p holds. If g is true, then p also holds after S, so $p \vee q$ holds.

23. (a)

$a = 2, n = 11$	p	q	i
initially	1	2	11
after first pass	2	4	5
after second pass	8	16	2
after third pass	8	256	1
after fourth pass	256·8	256^2	0

(b) We need to show that if $q^i p = a^n$ then $(new\ q)^{(new\ i)}(new\ p) = a^n$. If i is odd, then $i = 2 \cdot (new\ i) + 1$, new $q = q^2$ and new $p = p \cdot q$, so

$(new\ q)^{(new\ i)}(new\ p) = (q^2)^{(new\ i)} p \cdot q = q^{2 \cdot (new\ i) + 1} \cdot p = q^i \cdot p = a^n.$

If i is even, then $i = 2 \cdot (new\ i)$, new $q = q^2$ and new $p = $ old $p = p$, so

$$(\text{new } q)^{(\text{new } i)}(\text{new } p) = (q^2)^{(\text{new } i)} \cdot p = q^i \cdot p = a^n.$$

Since i runs through a decreasing sequence of nonnegative integers, eventually $i = 0$ and the algorithm exits the loop. At this point $p = a^n$.

24. (a) On exit $m = 0$, so $x^n = x^0 \cdot y = y$.

  (b) Yes; $x^{m-1} \cdot z = x^n = x^m \cdot y$ if and only if $z = x \cdot y$.

Section 4.2

Induction proofs should be written carefully and completely. These answers will serve only as guides, *not* as models.

1. Check the basis. For the inductive step, assume the equality holds for k. Then

$$\sum_{i=1}^{k+1} i^2 = \sum_{i=1}^{k} i^2 + (k+1)^2 = \frac{k(k+1)(2k+1)}{6} + (k+1)^2.$$

Some algebra shows that the right-hand side equals

$$\frac{(k+1)(k+2)(2k+3)}{6},$$

and so the equality holds for $k+1$ whenever it holds for k.

2. Let $p(n)$ be "$4 + 10 + 16 + \cdots + (6n - 2) = n(3n + 1)$." Then $p(1)$ is "$4 = 1 \cdot (3 + 1)$" which is true by inspection. Assume that $p(k)$ is true for some $k \in \mathbb{P}$. Then

$$4 + 10 + \cdots + (6k - 2) + ((6(k+1) - 2)$$
$$= k(3k + 1) + (6(k + 1) - 2) \qquad \text{[since } p(k) \text{ is assumed true]}$$
$$= 3k^2 + 7k + 4$$
$$= (k + 1)(3k + 4) \qquad \text{[algebra]}$$
$$= (k + 1)(3(k + 1) + 1),$$

so $p(k + 1)$ is true. That is, $p(k) \Rightarrow p(k + 1)$.

By induction, $p(n)$ is true for all $n \in \mathbb{P}$.

3. (a) Take $n = 37^{20}$ in Example 1.

  (b) Take $n = 37^4$ in Example 1.

  (c) By (a), (b) and Example 1, $(37^{500} - 37^{100}) + (37^{100} - 37^{20}) + (37^{20} - 37^4)$ is a multiple of 10.

  (d) Calculate $37^4 - 1 = 1,874,160$ directly or observe that $37^5 - 37 = 37 \cdot (37^4 - 1)$ is a multiple of 5 but 37 is not.

  (e) By (c) and (d), as in (c).

4. The algebra in the inductive step is

$$\frac{k}{4k + 1} + \frac{1}{(4k + 1)(4k + 5)} = \frac{k(4k + 5) + 1}{(4k + 1)(4k + 5)}$$
$$= \frac{4k^2 + 5k + 1}{(4k + 1)(4k + 5)} = \frac{(4k + 1)(k + 1)}{(4k + 1)(4k + 5)} = \frac{k + 1}{4(k + 1) + 1}.$$

5. The basis is "$s_0 = 2^0 a + (2^0 - 1)b$," which is true since $2^0 = 1$ and $s_0 = a$. Assume inductively that $s_k = 2^k a + (2^k - 1)b$ for some $k \in \mathbb{N}$. The algebra in the inductive step is

$$2 \cdot [2^k a + (2^k - 1)b] + b = 2^{k+1} a + 2^{k+1} b - 2b + b.$$

6. (a) 1, 4, 9, 16.
   (b) Let $p(n) = $ "$S = n^2$ at the start of the nth pass." For the inductive step use $k^2 + 2\sqrt{k^2} + 1 = (k + 1)^2$.

7. Show that $11^{k+1} - 4^{k+1} = 11 \cdot (11^k - 4^k) + 7 \cdot 4^k$. Imitate Example 2(d).

8. (a) $m = 4$, $p(k) = $ "$2^k < k!$."
   (b) The verification was given in Example 2(a).
   (c) No. To get $p(k)$ true after each pass, we need first to know that $p(m)$ is true on entry into the loop. It turns out that m has to be at least 20.

9. (a) Suppose that $\sum_{i=0}^{k} 2^i = 2^{k+1} - 1$ and $0 \le k$. Then

$$\sum_{i=0}^{k+1} 2^i = (\sum_{i=0}^{k} 2^i) + 2^{k+1} = 2^{k+1} - 1 + 2^{k+1} = 2^{k+2} - 1,$$

so the equation still holds for the new value of k.
   (b) Like (a).
   (c) Yes. $\sum_{i=0}^{0} 2^i = 1 = 2^1 - 1$ initially, so the loop never exits and the invariant is true for every value of k in $\mathbb{N}$.
   (d) No. The loop gets stuck at $k = 0$.

10. Basis: $2^2 = 4 > 3 = 2 + 1$. For the inductive step, if $k^2 > k + 1$ then $(k + 1)^2 = k^2 + 2k + 1 > k + 1 + 2k + 1 \ge k + 2$ [since $k \ge 0$].
    Simplest without induction: for $n \ge 2$ we have $n^2 \ge 2n = n + n > n + 1$.

11. (a) $1 + 3 + \cdots + (2n - 1) = n^2$.
    (b) For the inductive step $k^2 + [2(k + 1) - 1] = k^2 + 2k + 1 = (k + 1)^2$.

12. If $n \geq 6$ then $n^2 - 4n - 7 = n(n - 4) - 7 \geq 6 \cdot (6 - 4) - 7 = 5 > 0$. No induction required. The inequality fails for $n \leq 5$ by calculation.

13. (a) Assume $p(k)$ is true. Then $(k + 1)^2 + 5(k + 1) + 1 = (k^2 + 5k + 1) + (2k + 6)$. Since $k^2 + 5k + 1$ is even by assumption and $2k + 6$ is clearly even, $p(k + 1)$ is true.
(b) All propositions $p(n)$ are false. *Moral:* The basis of induction is crucial for mathematical induction. Also, see Exercises 8 and 9.

14. For the inductive step, if $\displaystyle\sum_{i=k}^{2k-1} (2i + 1) = 3k^2$, then

$$\sum_{i=k+1}^{2(k+1)-1} (2i + 1) = \sum_{i=k}^{2k-1} (2i +1) - [2k + 1] + [2 \cdot (2k) + 1] + [2 \cdot (2k + 1) + 1]$$

$$= 3k^2 - 2k - 1 + 4k + 1 + 4k + 2 + 1$$

$$= 3k^2 + 6k + 3 = 3(k + 1)^2.$$

Alternatively, we use Example 2(b): $\displaystyle\sum_{i=1}^{n} i = \tfrac{1}{2}n(n + 1)$. Thus

$$\sum_{i=n}^{2n-1} (2i + 1) = 2 \cdot \sum_{i=n}^{2n-1} i + \sum_{i=n}^{2n-1} 1$$

$$= 2\left\{ \sum_{i=1}^{2n-1} i - \sum_{i=1}^{n-1} i \right\} + [(2n - 1) - (n - 1)]$$

$$= 2\{\tfrac{1}{2}(2n - 1) \cdot 2n - \tfrac{1}{2}(n - 1) \cdot n\} + n$$

$$= 4n^2 - 2n - n^2 + n + n = 3n^2.$$

For a third solution, note that $1 + 3 + \cdots + (2n - 1) = \displaystyle\sum_{i=0}^{n-1} (2i + 1) = n^2$

by Exercise 11, and so $\displaystyle\sum_{i=n}^{2n-1} (2i + 1) = \sum_{i=0}^{2n-1} (2i + 1) - \sum_{i=0}^{n-1} (2i + 1) = (2n)^2 - n^2 = 3n^2$.

15. *Hint:* $5^{k+1} - 4(k + 1) - 1 = 5(5^k - 4k - 1) + 16k$.

16. If we use the formula $\displaystyle\sum_{i=1}^{n} i = \tfrac{1}{2}n(n + 1)$ from Example 2(b), the algebra for the inductive step becomes

$$\sum_{i=1}^{k+1} i^3 = \sum_{i=1}^{k} i^3 + (k + 1)^3 = [\sum_{i=1}^{k} i]^2 + (k + 1)^3$$

$$= [\tfrac{1}{2}k(k + 1)]^2 + (k + 1)^3 = \frac{k^2(k + 1)^2}{4} + \frac{4(k + 1)(k + 1)^2}{4}$$

$$= \frac{(k^2 + 4k + 4)(k + 1)^2}{4} = [\tfrac{1}{2}(k + 2)(k + 1)]^2 = [\sum_{i=1}^{k+1} i]^2.$$

17. *Hints:*

$$\frac{1}{n+2} + \cdots + \frac{1}{2n+2} = \left(\frac{1}{n+1} + \cdots + \frac{1}{2n}\right) + \left(\frac{1}{2n+1} + \frac{1}{2n+2} - \frac{1}{n+1}\right)$$

and

$$\frac{1}{2n+1} + \frac{1}{2n+2} - \frac{1}{n+1} = \frac{1}{2n+1} - \frac{1}{2n+2}.$$

Alternatively, to avoid induction, let $f(n) = \sum_{i=1}^{n} \frac{1}{i}$ and write both sides in terms of f. The

left-hand side is $f(2n) - f(n)$ and the right-hand side is

$1 + (1/2) + (1/3) + \cdots + (1/2n) - 2 \cdot [(1/2) + (1/4) + \cdots + (1/2n)]$

$= f(2n) - 2 \cdot \frac{1}{2} \cdot f(n)$.

18. (a) To prove $\sqrt{k} + \frac{1}{\sqrt{k+1}} \geq \sqrt{k+1}$, for the inductive step observe that $\sqrt{k}\sqrt{k+1} >$

$\sqrt{k \cdot k} = k$, so $\sqrt{k}\sqrt{k+1} + 1 > k + 1 = \sqrt{k+1}\sqrt{k+1}$. Divide by $\sqrt{k+1}$.

(b) To prove $2\sqrt{k} - 1 + \frac{1}{\sqrt{k+1}} \leq 2\sqrt{k+1} - 1$, observe that

$$k(k+1) = k^2 + k < k^2 + k + \frac{1}{4} = (k + \frac{1}{2})^2,$$

so $\sqrt{k}\sqrt{k+1} < k + \frac{1}{2}$, $2\sqrt{k}\sqrt{k+1} + 1 < 2k + 2$ and thus

$$2\sqrt{k} + \frac{1}{\sqrt{k+1}} < \frac{(2k+2)}{\sqrt{k+1}} = 2\sqrt{k+1}.$$

19. *Hints:* $5^{k+2} + 2 \cdot 3^{k+1} + 1 = 5(5^{k+1} + 2 \cdot 3^k + 1) - 4(3^k + 1)$. Show that $3^n + 1$ is always even.

20. For the basis, $8^2 + 9 = 73$. For the inductive step, $8^{k+3} + 9^{2k+3} = 8 \cdot 8^{k+2} + 8 \cdot 9^{2k+1} -$
$8 \cdot 9^{2k+1} + 81 \cdot 9^{2k+1} = 8 \cdot (8^{k+2} + 9^{2k+1}) + (81 - 8) \cdot 9^{2k+1}$.

21. Here p(n) is the proposition "$|\sin nx| \leq n|\sin x|$ for all $x \in \mathbb{R}$." Clearly p(1) holds. By algebra and trigonometry,

$|\sin (k + 1)x| = |\sin (kx + x)| = |\sin kx \cos x + \cos kx \sin x|$

$\leq |\sin kx| \cdot |\cos x| + |\cos kx| \cdot |\sin x| \leq |\sin kx| + |\sin x|$.

Now assume p(k) is true and show p(k+1) is true.

Section 4.3

1. (a) 1, 2, 1, 2, 1, 2, 1, 2, ... (b) {1, 2}.

2. 1, 1/2, 2/3, 3/5, 8/13.

3. (a) SEQ(n) = 3^n.

(b) (B) $\text{SEQ}(0) = 1$,

(R) $\text{SEQ}(n + 1) = 3 \cdot \text{SEQ}(n)$ for $n \geq 1$.

4. (a) $\text{SEQ}(0) = 2$ and $\text{SEQ}(n + 1) = \text{SEQ}(n)^2$ for $n \in \mathbb{N}$.

(b) $\text{SEQ}(0) = 2$ and $\text{SEQ}(n + 1) = 2^{\text{SEQ}(n)}$ for $n \in \mathbb{N}$.

5. No. It's okay up to $\text{SEQ}(100)$, but $\text{SEQ}(101)$ is not defined, since we cannot divide by zero. If, in (R), we restricted n to be ≤ 100, we would obtain a recursively defined *finite* sequence.

6. (a) $\text{SEQ}(9) = 315/128$.   (b) $\text{FIB}(11) = 144$.

7. (a) $1, 3, 8$.   (b) $s_n = 2s_{n-1} + 2s_{n-2}$ for $n \geq 2$.   (c) $s_3 = 22$, $s_4 = 60$.

8. (b) In the allowable words all b's precede all a's. The number of a's can be $0, 1, \ldots, n$, so $s_n = n + 1$ for $n \in \mathbb{N}$.

9. (a) $1, 1, 2, 4$.

(b) As in Example 5 we get $t_n = t_{n-1} + (2^{n-1} - t_{n-1}) = 2^{n-1}$ for $n \geq 1$, so no induction is required.

(c) Ours doesn't, since $t_0 \neq 1$.

10. (a) $1, 0, 1, 0, 1, 0, 1, 0, 1, 0, \ldots$.   (b) $\{0, 1\}$.

11. (a) $a_6 = a_5 + 2a_4 = a_4 + 2a_3 + 2a_4 = 3(a_3 + 2a_2) + 2a_3 =$
$5(a_2 + 2a_1) + 6a_2 = 11(a_1 + 2a_0) + 10a_1 = 11 \cdot 3 + 10 = 43$. This calculation uses only two intermediate value addresses at any given time. Other recursive calculations are possible that use more.

(b) Use induction. By definition a_n is odd for $n = 0$ and $n = 1$. Assume that a_n is odd. Since $2a_{n-1}$ is even, $a_n + 2a_{n-1}$ is odd, but this is exactly a_{n+1}.

12. By computation, $M_1 \cdot M_1 = M_2$. Since $\text{FIB}(k + 1) = \text{FIB}(k) + \text{FIB}(k - 1)$ for $k \geq 1$,

$$M_1 \cdot M_n = \begin{bmatrix} 1 & 1 \\ 1 & 0 \end{bmatrix} \cdot \begin{bmatrix} \text{FIB}(n) & \text{FIB}(n - 1) \\ \text{FIB}(n - 1) & \text{FIB}(n - 2) \end{bmatrix} =$$

$$\begin{bmatrix} \text{FIB}(n) + \text{FIB}(n - 1) & \text{FIB}(n - 1) + \text{FIB}(n - 2) \\ \text{FIB}(n) & \text{FIB}(n - 1) \end{bmatrix} = M_{n+1} \text{ for } n \geq 2.$$

13. Follow the hint. Suppose $S \neq \emptyset$. By the Well-Ordering Principle, S has a smallest member, say m. Since $s_0 = 1 = \text{FIB}(1)$ and $s_1 = 2 = \text{FIB}(2)$, $m \geq 2$. Then $s_m = s_{m-1} + s_{m-2} = \text{FIB}(m - 1 + 1) + \text{FIB}(m - 2 + 1)$ [since m was the smallest bad guy] $= \text{FIB}(m + 1)$ [by recursive definition of FIB in Example 3(a)], contrary to $m \in S$. Thus $S = \emptyset$.

14. (a) 1, 1, 3, 7, 17, 41.
    (b) The first two terms are odd and if b_{n-2} is odd then b_n is an odd integer. A proof using our present version of induction is awkward, since if $p(n)$ is "b_n is odd," then $p(n) \rightarrow p(n + 2)$ is fairly clear but $p(n) \rightarrow p(n + 1)$ is not. We will cover a more general form of induction in § 4.5 that handles this situation. See Example 2 of that section.

    If a proof is desired at this point, we can use the Well-Ordering Principle. Let $S = \{n \in \mathbb{N} : b_n \text{ is even}\}$ and suppose that S is not empty. Then S has a smallest member, say m. Now $m \neq 0$ and $m \neq 1$ since b_0 and b_1 are odd integers. Thus $m \geq 2$ so $m - 2$ is in $\mathbb{N}$. Since b_m is even, so is $b_{m-2} = b_m - 2b_{m-1}$, contradicting the choice of m. Hence S must be empty; i.e., b_n is odd for each $n \in \mathbb{N}$.

15. $\text{SEQ}(n) = 2^{n-1}$ for $n \geq 1$.

16. (a) 0, 1, 2, 1, 0, 1, 2, 1, 0, 1, ...   (b) $\{0, 1, 2\}$.

17. (a) $A(1) = 1$. $A(n) = n \cdot A(n - 1)$.
    (b) $A(6) = 6 \cdot A(5) = 6 \cdot 5 \cdot A(4) = \cdots = 6! = 720$.
    (c) Yes.

18 (a) $B(1) = 1$. $B(n) = \dfrac{(2n)(2n - 1)}{2} \cdot B(n - 1)$ for $n \geq 2$.
    (b) $B(3) = \dfrac{6 \cdot 5}{2} \cdot B(2) = \dfrac{6 \cdot 5}{2} \cdot \dfrac{4 \cdot 3}{2} \cdot 1 = 90$.
    (c) $B(5) = 113,400$ [just for 10 children!].
    (d) $B(n) = (2n)!/2^n$.

19. (a) $\{1, 110, 1200\}$.   (b) $\{1, 2, 55, 120, 650\}$.

20. (a) $\text{SUM}(1) = a_1$ and $\text{SUM}(n + 1) = \text{SUM}(n) + a_{n+1}$ for $n \in \mathbb{P}$.
    (b) $\text{SUM}(0) = 0$ and $\text{SUM}(n + 1) = \text{SUM}(n) + a_{n+1}$ for $n \in \mathbb{N}$. The "empty sum" is 0.
    [Think of default content of a register.]

21. (a) (B) UNION(1) = A_1,

   (R) UNION(n) = $A_n \cup$ UNION(n - 1) for $n \geq 2$.

   (b) The "empty union" is $\emptyset$.

   (c) (B) INTER(1) = A_1,

   (R) INTER(n + 1) = $A_{n+1} \cap$ INTER(n) for $n \in \mathbb{P}$

   [or INTER(n) = $A_n \cap$ INTER(n - 1) for $n \geq 2$].

   (d) Empty intersection should be the universe, in this case S.

22. Let p(n) be "$x \in$ SYM(n) if and only if $\{k : x \in A_k$ and $k \leq n\}$ has an odd number of elements." Then p(1) is "$x \in A_1$ if and only if $\{k : x \in A_k$ and $k \leq 1\}$ has an odd number of elements," by (B). Since $x \in A_k$ and $k \leq 1$ if and only if $k = 1$ and $x \in A_1$, p(1) is true.

   Assume inductively that p(n) is true for some $n \in \mathbb{P}$. Consider $x \in$ SYM(n + 1) = $A_{n+1} \oplus$ SYM(n). If $x \in A_{n+1}$, then $x \notin$ SYM(n), so by p(n) the set $\{k : x \in A_k$ and $k \leq n\}$ has an even number of elements, say 2m. Hence the set $\{k : x \in A_k$ and $k \leq n + 1\}$ has $2m + 1$ elements [including $n + 1$]; i.e., it has an odd number of elements. If $x \in$ SYM(n), then $x \notin A_{n+1}$ and $\{k : x \in A_k$ and $k \leq n + 1\}$ = $\{k : x \in A_k$ and $k \leq n\}$, which has an odd number of elements, by p(n). We have shown that if $x \in$ SYM(n + 1) then $\{k : x \in A_k$ and $k \leq n + 1\}$ has an odd number of elements. If $x \in S \setminus$ SYM(n + 1) then either $x \notin A_{n+1} \cup$ SYM(n) or $x \in A_{n+1} \cap$ SYM(n). In the first case $\{k : x \in A_k$ and $k \leq n + 1\}$ = $\{k : x \in A_k$ and $k \leq n\}$, which has an even number of elements by p(n). In the second case the set $\{k : x \in A_k$ and $k \leq n + 1\}$ = $\{n + 1\} \cup \{k : x \in A_k$ and $k \leq n\}$, which has an even number of elements by p(n). Thus if $x \in S \setminus$ SYM(n + 1) then $\{k : x \in A_k$ and $k \leq n + 1\}$ has an even number of elements.

   This fact and the result of the last paragraph show that p(n + 1) is true whenever p(n) is true. It follows by induction that p(n) is true for every $n \in \mathbb{P}$.

Section 4.4

1. $s_n = 3 \cdot (-2)^n$ for $n \in \mathbb{N}$.

2. (a) $s_{2n} = 4^n$ and $s_{2n + 1} = 4^n$ for $n \in \mathbb{N}$.

   (b) $s_{2n} = 4^n$ and $s_{2n + 1} = 2 \cdot 4^n$ for $n \in \mathbb{N}$; i.e., $s_n = 2^n$ for all $n \in \mathbb{N}$.

3. We prove this by induction. $s_n = a^n \cdot s_0$ holds for $n = 0$ because $a^0 = 1$. If it holds for some n, then $s_{n+1} = as_n = a(a^n \cdot s_0) = a^{n+1} \cdot s_0$ and so the result holds for $n + 1$.

4. $s_0 = 2^{0+1} + (-1)^0 = 3$. $s_1 = 2^{1+1} + (-1)^1 = 3$. For $n \geq 2$,

$$s_{n-1} + 2s_{n-2} = 2^{(n-1)+1} + (-1)^{n-1} + 2[2^{(n-2)+1} + (-1)^{n-2}]$$
$$= 2^n - (-1)^n + 2^n + 2(-1)^n$$
$$= 2^{n+1} + (-1)^n = s_n.$$

5. $s_0 = 3^0 - 2 \cdot 0 \cdot 3^0 = 1.$ $s_1 = 3^1 - 2 \cdot 1 \cdot 3^1 = -3.$ For $n \geq 2,$
$$6s_{n-1} - 9s_{n-2} = 6[3^{n-1} - 2(n-1) \cdot 3^{n-1}] - 9[3^{n-2} - 2(n-2) \cdot 3^{n-2}]$$
$$= 2[3^n - 2(n-1) \cdot 3^n] - [3^n - 2(n-2) \cdot 3^n]$$
$$= 3^n[2 - 4(n-1) - 1 + 2(n-2)]$$
$$= 3^n[1 - 2n] = s_n.$$

6. Calculate $\dfrac{1}{\sqrt{5}}\left[\left(\dfrac{1+\sqrt{5}}{2}\right)^6 - \left(\dfrac{1-\sqrt{5}}{2}\right)^6\right].$

7. This time $c_1 = 3$ and $c_2 = 0$ and so $s_n = 3 \cdot 2^n$ for $n \in \mathbb{N}.$

8. $c_1 = 0,$ $c_2 = 3$ and so $s_n = 3 \cdot (-1)^n$ for $n \in \mathbb{N}.$

9. Solve $1 = c_1 + c_2$ and $2 = c_1 r_1 + c_2 r_2$ for c_1 and c_2 to obtain $c_1 = (1 + r_1)/\sqrt{5}$ and $c_2 = -(1 + r_2)/\sqrt{5}.$ Hence
$$s_n = \frac{1}{\sqrt{5}}(r_1^n + r_1^{n+1} - r_2^n - r_2^{n+1}) \quad \text{for all } n,$$

where r_1, r_2 are as in Example 3.

10. (a) 3, 4, 7, 11, 18.
(b) Solve $2 = c_1 + c_2$ and $1 = c_1 r_1 + c_2 r_2$ to obtain $c_1 = c_2 = 1.$ So $s_n = r_1^n + r_2^n$ for $n \in \mathbb{N},$ where r_1, r_2 are as in Example 3.

11. (a) $r_1 = -3,$ $r_2 = 2,$ $c_1 = c_2 = 1$ and so $s_n = (-3)^n + 2^n$ for $n \in \mathbb{N}.$
(b) $s_n = 2 \cdot 5^n$ for $n \in \mathbb{N}.$
(c) Here the characteristic equation has one solution $r = 2.$ Then $c_1 = 1$ and $c_2 = 3$ and so $s_n = 2^n + 3n \cdot 2^n$ for $n \in \mathbb{N}.$
(d) Here $r_1 = 2,$ $r_2 = 3.$ Solve $c = c_1 + c_2$ and $d = 2c_1 + 3c_2$ to obtain $c_1 = 3c - d$ and $c_2 = d - 2c.$ So $s_n = (3c - d) \cdot 2^n + (d - 2c) \cdot 3^n$ for $n \in \mathbb{N}.$
(e) $s_{2n} = 1,$ $s_{2n+1} = 4$ for all $n \in \mathbb{N}.$
(f) $s_{2n} = 3^n$ and $s_{2n+1} = 2 \cdot 3^n$ for $n \in \mathbb{N}.$
(g) $s_n = (-3)^n$ for $n \in \mathbb{N}.$
(h) $s_n = -(1/4)(-3)^n + (5/4)$ for $n \in \mathbb{N}.$

12. The characteristic equation is $x^2 - b = 0$ and the solutions are $r_1 = \sqrt{b}$ and $r_2 = -\sqrt{b}$.
Solving $s_0 = c_1 + c_2$ and $s_1 = c_1\sqrt{b} - c_2\sqrt{b}$ yields $c_1 = \frac{1}{2}(s_0 + s_1/\sqrt{b})$, $c_2 = \frac{1}{2}(s_0 - s_1/\sqrt{b})$.

Now

$$s_n = \frac{1}{2}(s_0 + s_1/\sqrt{b})(\sqrt{b})^n + \frac{1}{2}(s_0 - s_1/\sqrt{b})(-\sqrt{b})^n$$
$$= (\sqrt{b})^n \cdot \frac{1}{2}[s_0 + s_1/\sqrt{b} + (-1)^n(s_0 - s_1/\sqrt{b})].$$

So

$$s_{2n} = (\sqrt{b})^{2n} \cdot \frac{1}{2}[s_0 + s_1/\sqrt{b} + s_0 - s_1/\sqrt{b}] = b^n \cdot s_0$$

and

$$s_{2n+1} = (\sqrt{b})^{2n+1} \cdot \frac{1}{2}[s_0 + s_1/\sqrt{b} - s_0 + s_1/\sqrt{b}] = b^n \cdot s_1.$$

13. (a) $s_{2^m} = 2^m + 3 \cdot (2^m - 1) = 2^{m+2} - 3.$ (b) $s_{2^m} = 3 \cdot 2^m.$

   (c) $s_{2^m} = \frac{5}{2} \cdot 2^m \cdot m.$ (d) $s_{2^m} = 2^{m+1} + 3 \cdot (2^m - 1) + \frac{5}{2} \cdot 2^m \cdot m.$

   (e) $s_{2^m} = 7 - 6 \cdot 2^m.$ (f) $s_{2^m} = 7 - 2^{m+1}.$

   (g) $s_{2^m} = (6 - m)2^{m-1}.$ (h) $s_{2^m} = (10 - 7m)2^{m-1} - 5.$

14. (a) $s_{2^m} \le 2^m \cdot 7 + 2^m - 1 = 2^{m+3} - 1.$ (b) $s_{2^m} \le 2^m \cdot 7 + m \cdot 2^m = (7 + m)2^m.$

15. $s_{2^m} = 2^m[s_1 + \frac{1}{2}(2^m - 1)]$. Verify that this formula satisfies $s_{2^0} = s_1$ and
$s_{2^{m+1}} = 2s_{2^m} + (2^m)^2.$

16. The inductive step is
$$s_{2^{m+1}} = 2s_{2^m} + A + B \cdot 2^m$$
$$= 2[2^m s_1 + (2^m - 1)A + \frac{B}{2} \cdot 2^m \cdot m] + A + B \cdot 2^m$$
$$= 2^{m+1}s_1 + [2 \cdot (2^m - 1) + 1] \cdot A + [2^m \cdot m + 2^m] \cdot B$$
$$= 2^{m+1}s_1 + [2^{m+1} - 1] \cdot A + \frac{B}{2} \cdot 2^{m+1}(m+1).$$

17. (a) $t_{2^m} = b^m t_1 + b^{m-1} \cdot \sum_{i=0}^{m-1} \frac{f(2^i)}{b^i}.$

   (b) $t_{3^m} = 3^m t_1 + 3^{m-1} \cdot \sum_{i=0}^{m-1} \frac{f(3^i)}{3^i}.$

Section 4.5

1. The First Principle is adequate for this. For the inductive step, use the identity
$4n^2 - n + 8(n + 1) - 5 = 4n^2 + 7n + 3 = 4(n + 1)^2 - (n + 1).$

2. The First Principle applies in both parts.

3. Show that $n^5 - n$ is always even. Then use the identity $(n + 1)^5 = n^5 + 5n^4 + 10n^3 + 10n^2 + 5n + 1$ [from the binomial theorem]. More detail: The difference $n^5 - n$ is even since both n and n^5 are even if n is even and both are odd if n is odd. Use the First Principle of induction to show that $n^5 - n$ is always divisible by 5. If this is true for n, then this is true for $n + 1$ because $(n + 1)^5 - (n + 1) = n^5 - n + 5(n^4 + 2n^3 + 2n^2 + n)$.

4. (a) $b_6 = 2b_5 + b_4 = 2 \cdot 41 + 17 = 99$.
   (b) $a_9 = 9$. We omit the step-by-step computations of a_3, a_4, etc.

5. Yes. The oddness of a_n depends only on the oddness of a_{n-1}, since $2a_{n-2}$ is even whether a_{n-2} is odd or not.

6. (b) $a_n = 2^n$ for $n \in \mathbb{N}$.
   (c) Use the general Second Principle with $m = 0$, $\ell = 1$. Let $p(n) =$ "$a_n = 2^n$." By definition $p(0)$ and $p(1)$ are true. Consider $n \geq 2$ and assume that $a_k = 2^k$ for all k satisfying $0 \leq k < n$. Then
   $$a_n = \frac{a_{n-1}^2}{a_{n-2}} = \frac{(2^{n-1})^2}{2^{n-2}} = 2^{2n-2-n+2} = 2^n.$$
   By the Second Principle $a_n = 2^n$ for every $n \in \mathbb{N}$.

7. (b) $a_n = 1$ for all $n \in \mathbb{N}$.
   (c) The basis needs to be checked for $n = 0$ and $n = 1$. For the inductive step, consider $n \geq 2$ and assume $a_k = 1$ for $0 \leq k < n$. Then $a_n = \frac{a_{n-1}^2 + a_{n-2}}{a_{n-1} + a_{n-2}} = \frac{1^2 + 1}{1 + 1} = 1$. This completes the inductive step, and so $a_n = 1$ for all $n \in \mathbb{N}$ by the Second Principle of Induction.

8. (b) $a_n = n + 1$ for $n \in \mathbb{N}$.
   (c) Use the general Second Principle with $m = 0$, $\ell = 1$. Check for $n = 0, 1$. For the inductive step
   $$a_n = \frac{a_{n-1}^2 - 1}{a_{n-2}} = \frac{(n - 1 + 1)^2 - 1}{(n - 2 + 1)} = \frac{n^2 - 1}{n - 1} = n + 1.$$

9. (b) $a_n = n^2$ for all $n \in \mathbb{N}$.
   (c) The basis needs to be checked for $n = 0$ and $n = 1$. For the inductive step, consider $n \geq 2$ and assume that $a_k = k^2$ for $0 \leq k < n$. To complete the inductive step, note that
   $$a_n = \tfrac{1}{4}(a_{n-1} - a_{n-2} + 3)^2 = \tfrac{1}{4}[(n - 1)^2 - (n - 2)^2 + 3]^2 = \tfrac{1}{4}[2n]^2 = n^2.$$

10. (a) $a_3 = 4$, $a_4 = 7$, $a_5 = 10$, $a_6 = 15$, $a_7 = 24$.

(b) Check for $n = 1, 2, 3$. For the inductive step for $n \geq 4$,
$$a_n = a_{n-2} + 2a_{n-3} > \left(\tfrac{3}{2}\right)^{n-2} + 2\left(\tfrac{3}{2}\right)^{n-3} = \left(\tfrac{3}{2}\right)^n\left[\left(\tfrac{2}{3}\right)^2 + 2\left(\tfrac{2}{3}\right)^3\right] = \left(\tfrac{3}{2}\right)^n \cdot \tfrac{28}{27} > \left(\tfrac{3}{2}\right)^n.$$

11. (b) The basis needs to be checked for $n = 0, 1$ and 2. For the inductive step, consider $n \geq 3$ and assume that a_k is odd for $0 \leq k < n$. Then $a_{n-1}, a_{n-2}, a_{n-3}$ are all odd. Since the sum of three odd integers is odd [if not obvious, prove it], a_n is also odd.

(c) Since the inequality is claimed for $n \geq 1$ and since you will want to use the identity $a_n = a_{n-1} + a_{n-2} + a_{n-3}$ in the inductive step, you will need $n - 3 \geq 1$ in the inductive step. So check the basis for $n = 1, 2$ and 3. For the inductive step, consider $n \geq 4$ and assume that $a_k \leq 2^{k-1}$ for $1 \leq k < n$. To complete the inductive step, note that
$$a_n = a_{n-1} + a_{n-2} + a_{n-3} < 2^{n-2} + 2^{n-3} + 2^{n-4} = \tfrac{7}{8} \cdot 2^{n-1} < 2^{n-1}.$$

12. (a) $a_3 = 11$, $a_4 = 21$, $a_5 = 43$, $a_6 = 85$, $a_7 = 171$.

(b) Check for $n = 1, 2, 3$. For the inductive step with $n \geq 4$,
$$a_n = 3a_{n-2} + 2a_{n-3} > 3 \cdot 2^{n-2} + 2 \cdot 2^{n-3} = 2^n.$$

(c) Check for $n = 1, 2, 3$. For $n \geq 4$, $a_n = 3a_{n-2} + 2a_{n-3} < 3 \cdot 2^{n-1} + 2 \cdot 2^{n-2} = 2^{n+1}$.

(d) Check for $n = 1, 2, 3$. For $n \geq 4$,
$$a_n = 3a_{n-2} + 2a_{n-3} = 3 \cdot [2a_{n-3} + (-1)^{n-3}] + 2 \cdot [2a_{n-4} + (-1)^{n-4}]$$
$$= 2 \cdot [3a_{n-3} + 2a_{n-4}] + (-1)^{n-3} \cdot [3 - 2] = 2a_{n-1} + (-1)^{n-1}.$$

13. (a) 2, 3, 4, 6.

(b) The inequality must be checked for $n = 3, 4$ and 5 before applying the Second Principle of Mathematical Induction to $b_n = b_{n-1} + b_{n-3}$. For the inductive step, consider $n \geq 6$ and assume $b_k \geq 2b_{k-2}$ for $3 \leq k < n$. Then
$$b_n = b_{n-1} + b_{n-3} \geq 2b_{n-3} + 2b_{n-5} = 2b_{n-2}.$$

(c) The inequality must be checked for $n = 2, 3$ and 4. Then use the Second Principle of Mathematical Induction and part (b). For the inductive step, consider $n \geq 5$ and assume $b_k \geq (\sqrt{2})^{k-2}$ for $2 \leq k < n$. Then
$$b_n = b_{n-1} + b_{n-3} \geq 2b_{n-3} + b_{n-3} = 3b_{n-3} \geq 3(\sqrt{2})^{n-5} > (\sqrt{2})^3(\sqrt{2})^{n-5} = (\sqrt{2})^{n-2}.$$
Note that $3 > (\sqrt{2})^3 \approx 2.828$. This can also be proved without using part (b):
$$b_n \geq (\sqrt{2})^{n-3} + (\sqrt{2})^{n-5} = (\sqrt{2})^{n-2} \cdot \left[\tfrac{1}{\sqrt{2}} + \tfrac{1}{2^{3/2}}\right] > (\sqrt{2})^{n-2}.$$

14. Check for $n = 1, 2, 3$. For the induction step, consider $n \geq 4$ and assume $b_k \leq \left(\tfrac{3}{2}\right)^{k-1}$ for $1 \leq k < n$. Then
$$b_n = b_{n-1} + b_{n-3} \leq \left(\tfrac{3}{2}\right)^{n-2} + \left(\tfrac{3}{2}\right)^{n-4} = \left(\tfrac{3}{2}\right)^{n-1} \cdot \left[\left(\tfrac{2}{3}\right) + \left(\tfrac{2}{3}\right)^3\right] = \left(\tfrac{3}{2}\right)^{n-1} \cdot \tfrac{26}{27} < \left(\tfrac{3}{2}\right)^{n-1}.$$

15. Check for $n = 0$ and 1 before applying induction. It may be simpler to prove "$SEQ(n) \le 1$ for all n" separately from "$SEQ(n) \ge 0$ for all n". For example, assume that $n \ge 2$ and that $SEQ(k) \le 1$ for $0 \le k < n$. Then

$$SEQ(n) = (1/n){*}SEQ(n-1) + ((n-1)/n){*}SEQ(n-2) \le (1/n) + ((n-1)/n) = 1.$$

The proof that $SEQ(n) \ge 0$ for $n \ge 0$ is almost the same.

16. First note that $SEQ(n + 1) = \sum\limits_{i=0}^{n} SEQ(i) = SEQ(n) + \sum\limits_{i=0}^{n-1} SEQ(i) = SEQ(n) + SEQ(n) = 2{\cdot}SEQ(n).$

Now the First Principle is sufficient to prove $SEQ(n) = 2^{n-1}$ for $n \ge 1$. For the inductive step, $SEQ(n + 1) = 2{\cdot}SEQ(n) = 2{\cdot}2^{n-1} = 2^n.$

17. The First Principle is enough. Use (R) to check for $n = 2$. For the inductive step from n to $n + 1$,

$$FIB(n+1) = FIB(n) + FIB(n-1) = 1 + \sum\limits_{k=0}^{n-2} FIB(k) + FIB(n-1) = 1 + \sum\limits_{k=0}^{n-1} FIB(k).$$

18. (a) $1, 3, 4, 7, 11, 18$, etc.

(b) First check for $n = 2$ and 3. Then apply the Second Principle of Induction:

$$
\begin{aligned}
LUC(n) &= LUC(n - 1) + LUC(n - 2) \\
&= [FIB(n - 1) + FIB(n - 3)] + [FIB(n - 2) + FIB(n - 4)] \\
&= [FIB(n - 1) + FIB(n - 2)] + [FIB(n - 3) + FIB(n - 4)] \\
&= FIB(n) + FIB(n - 2).
\end{aligned}
$$

Explanations should be supplied.

19. For $n > 0$, let $L(n)$ be the largest integer 2^k with $2^k \le n$. Show that $L(n) = T(n)$ for all n by showing first that $L(\lfloor n/2 \rfloor) = L(n/2)$ for $n \ge 2$ and then using the Second Principle of Induction.

20. (a) In fact, $T(n) \le n$ for all n, by Exercise 19 or by a proof using the Second Principle of Induction.

(b) There are lots of proofs. To show that $Q(n) \le n^2$ by the Second Principle of Induction, observe in the proof of the inductive step that $n \le n^2/2$ for $n \ge 2$.

(c) Show that $Q(n) \le 2n \log n$ for $n \ge 2$. The inductive step is

$$Q(n) = 2{\cdot}Q(\lfloor n/2 \rfloor) + n \le 2{\cdot}2\lfloor n/2 \rfloor \log \lfloor n/2 \rfloor + n \le 2{\cdot}\frac{2n}{2}{\cdot}\log (n/2) + n =$$

$2n(\log n - 1) + n = 2n \log n - n < 2n \log n.$

21. Show that $S(n) \le n$ for every n by the Second Principle of Induction.

Section 4.6

1. (a) 20.   (c) 1.   (e) 4.   (g) 6.
   (b) 10.   (d) 20.   (f) 20.   (h) 1.

2. (a) 17.   (b) 1.   (c) 17.   (d) 170.   (e) 17.   (f) 170.

3. (a) (20,14), (14,6), (6,2), (2,0); gcd = 2.
   (b) (20,7), (7,6), (6,1), (1,0); gcd = 1.
   (c) (20,30), (30,20), (20,10), (10,0); gcd = 10.
   (d) (2000,987), (987,26), (26,25), (25,1), (1,0); gcd = 1.

4. (a) (30,30), (30,0); gcd = 30.   (b) (30,10), (10,0); gcd = 10.
   (c) (30,60), (60,30), (30,0); gcd = 30.
   (d) (3000,999), (999,3), (3,0); gcd = 3.

5. (a) $\gcd(20,14) = 2$, $s = -2$, $t = 3$.

a	q	s	t
20		1	0
14	1	0	1
6	2	1	-1
2	3	-2	3
0			

(b) $\gcd(72,17) = 1$, $s = -4$, $t = 17$.

a	q	s	t
72		1	0
17	4	0	1
4	4	1	-4
1	4	-4	17
0			

(c) $\gcd(20,30) = 10$, $s = -1$, $t = 1$.

a	q	s	t
20		1	0
30	0	0	1
20	1	1	0
10	2	-1	1
0			

(d) $\gcd(320,30) = 10$, $s = -1$, $t = 11$.

a	q	s	t
320		1	0
30	10	0	1
20	1	1	-10
10	2	-1	11
0			

6. (a) $\gcd(14259,3521) = 7$, $s = 161$, $t = -652$.

a	q	s	t
14259		1	0
3521	4	0	1
175	20	1	-4
21	8	-20	81
7	3	161	-652
0			

(b) $\gcd(8359,9373) = 13$, $s = 342$, $t = -305$.

a	q	s	t
8359		1	0
9373	0	0	1
8359	1	1	0
1014	8	-1	1
247	4	9	-8
26	9	-37	33
13	2	342	-305
0			

7. (a) $x = 21$ [$\equiv -5 \pmod{26}$].   (b) $x = 19$ [$\equiv -7 \pmod{26}$].

(c) No solution exists, because 4 and 26 are not relatively prime.

(d) $x = 3$.   (e) $x = 23$ [$\equiv -3 \pmod{26}$].

(f) No solution exists, because 13 and 26 are not relatively prime.

8. (a) $x = 5$.   (b) $x = 11$.

(c) No solution exists, because 4 and 24 are not relatively prime.

(d) No solution exists, because 9 and 24 are not relatively prime.

(e) $x = 17$.   (f) $x = 13$.

9. (a) $x \equiv 5 \pmod{13}$.   (c) Same as (a), since $99 \equiv 8 \pmod{13}$.

(b) $x \equiv 5 \cdot 4 \equiv 7 \pmod{13}$.   (d) $x \equiv 5 \cdot 5 \equiv 12 \pmod{13}$.

10. (a) $x \equiv -163 \equiv 480 \pmod{643}$.

(b) $x \equiv 507 \pmod{2000}$.   (c) $x \equiv 111 \pmod{788}$.

(d) $x \equiv 24 \cdot (-232) \equiv -5568 \equiv 1020 \pmod{1647}$.

11. (a) $x = 99$.

Details: $x = 99y$, so $8y \equiv 99y \equiv 8 \pmod{13}$. Cancel a factor of 8 to get $y \equiv 1 \pmod{13}$. Taking $y = 1$ gives $x = 99$.

(b) $x = 65$.

Details: $x = 13y$ gives $13y \equiv 65 \pmod{99}$. Cancel a factor of 13 to get $y \equiv 5 \pmod{99}$. Taking $y = 5$ gives $x = 65$.

(c) $x = 164$. It is not an accident that the answers for parts (a) and (b) sum to the answer for (c). If x solves (a) and x' solves (b), then $x + x'$ solves (c).

12. (a) $x = 1, 1288$.　　　　　　　　(b) $x = 1286$ [$\equiv -1 \pmod{1287}$].
　　(c) $x = 1285$ [$\equiv -2 \pmod{1287}$].　　(d) $x = 1284$ [$\equiv -3 \pmod{1287}$].

13. Assume that $a = s \cdot m + t \cdot n$ and $a' = s' \cdot m + t' \cdot n$ at the start of the loop. The equation $a' = s' \cdot m + t' \cdot n$ becomes $a_{next} = s_{next} \cdot m + t_{next} \cdot n$ at the end, and $a'_{next} = a' - q \cdot a = s' \cdot m + t' \cdot n - q \cdot s \cdot m - q \cdot t \cdot n = (s' - q \cdot s) \cdot m + (t' - q \cdot t) \cdot n = s'_{next} \cdot m + t'_{next} \cdot n$ at the end.

14. By Theorem 3, $\gcd(m,n) = s \cdot m + t \cdot n$ for some s and t, so $\gcd(m,n)$ is of this form. Suppose that $e = a \cdot m + b \cdot n$ with $0 < e \le \gcd(m,n)$. Since $\gcd(m,n)$ divides both m and n, it divides e, just as in the Corollary to Theorem 3. Hence $\gcd(m,n) \le e$.

15. (a) $1 = s \cdot (m/d) + t \cdot (n/d)$ for some integers s and t. Apply Exercise 14 to m/d and n/d in place of m and n. A longer proof can be based on prime factorization.
    (b) $k = s't - st'$ works. Note that $d \cdot s = ss' \cdot m + st' \cdot n$ and $d \cdot s' = s's \cdot m + s't \cdot n$; subtract.
    (c) Let $x = s \cdot a/d$.

16. (a) $s' \cdot m + t' \cdot n = s \cdot m + k \cdot (n/d) \cdot m + t \cdot n - k \cdot (m/d) \cdot n = s \cdot m + t \cdot n = d$.
    (b) Let $s' = s \text{ MOD } \frac{n}{d}$ and $t' = t + (s \text{ DIV } \frac{n}{d}) \cdot \frac{m}{d}$, so that $t' \cdot n = t \cdot n + m \cdot (s \text{ DIV } \frac{n}{d}) \cdot \frac{n}{d} = t \cdot n + m \cdot (s - s \text{ MOD } \frac{n}{d})$.

17. (a) Check for $\ell = 1$. For the inductive step it suffices to show that if $a = m = \text{FIB}(\ell + 2)$ and $b = n = \text{FIB}(\ell + 1)$, $\ell \ge 1$, then after the first pass of the while loop, $a = \text{FIB}(\ell + 1)$ and $b = \text{FIB}(\ell)$. For then, by the inductive hypothesis, exactly ℓ more passes would be needed before terminating the algorithm. By the definition of (a,b) in the while loop, it suffices to show that $\text{FIB}(\ell + 2) \text{ MOD FIB}(\ell + 1) = \text{FIB}(\ell)$. But $\text{FIB}(\ell + 2) \text{ MOD FIB}(\ell + 1)$ $= [\text{FIB}(\ell + 1) + \text{FIB}(\ell)] \text{ MOD FIB}(\ell + 1) =$ $\text{FIB}(\ell) \text{ MOD FIB}(\ell + 1) = \text{FIB}(\ell)$, since $\text{FIB}(\ell) < \text{FIB}(\ell + 1)$ for $\ell \ge 1$.
    (b) Use induction on k. Check for $k = 3$. For the inductive step
    $$\log_2 \text{FIB}(k + 1) = \log_2 (\text{FIB}(k) + \text{FIB}(k - 1)) \le \log_2 (2\text{FIB}(k)) = 1 + \log_2 \text{FIB}(k).$$

[This estimate is far from best possible. The Fibonacci numbers are the worst case for the Euclidean algorithm.]
    (c) By part (a), GCD makes ℓ passes through the loop. By part (b), $\log_2 (m + n) = \log_2 \text{FIB}(\ell + 2) \le \ell$ if $\ell \ge 3$.

CHAPTER 5

Most of this chapter is about formulas that count sets without actually listing their members. A section on probability seems to fit naturally here, since lots of elementary probability questions can be answered by counting.

Section 5.1 lays out some basic terminology and simple techniques. You might want to draw a couple more examples of decision trees in the style of Figure 1 and then ask students what the tree for choosing a 5-card poker hand would look like. Although the tree is a good conceptual tool, we don't want to list all its leaves; we just want to know how many there are. The tree helps bring Product Rule (b) to life.

The account of probability in § 5.2 is just meant to be an introduction; Chapter 9 contains an extended treatment. The first dozen exercises give a sense of what we really expect students to be able to do here, but it doesn't hurt to assign some of the more challenging ones as well.

Section 5.3 presents two new methods and proves the binomial theorem. The Inclusion-Exclusion Principle is a little awkward to write out as a formula, but it is easy to illustrate and to understand. The formula for placing objects in boxes can be remembered more easily if its proof is understood. The examples in the text are more complicated than the exercises, on purpose. Students should be encouraged to follow arguments they would perhaps not arrive at themselves. They need practice reading with understanding, and the examples illustrate extensions of the basic techniques. Moreover, as we point out in Example 4(b), it is very easy to ask very hard combinatorial questions, so we cannot expect the exercises to come close to covering all situations that arise naturally. We don't assign Exercise 6, but do call attention to it in class.

Section 5.4 contains still more counting tools. It is not an accident that the two main formulas we get are the same. Students need to see the connection, as illustrated in Example 3(b). Exercise 5 should be assigned or done in class.

In a sense, the methods in §§ 5.3 and 5.4 are like techniques for integration in a calculus class. They don't handle everything, but they do provide practical tools. In the same way, the Pigeon-Hole Principle in § 5.5 is like the Mean Value Theorem. It says that something must occur, but it doesn't help find the occurrence. Section 5.5 can be covered in one day, though here again the text is trickier than many of the exercises. Really weak students won't be able to do

70

much with any of the problems. Do Example 4 or Example 5 in class - Ross likes 4 and Wright likes 5 - and urge the students to read the other one.

Section 5.1

1. (a) 56.   (b) 1.   (c) 56.   (d) 1326.   (e) 1.   (f) 52.

2. (a) 14, 4, 10.   (b) $2^{10} = 1024$.   (c) $\binom{10}{4} = 210$.   (d) $\binom{5}{3} \cdot 5 = 50$.

3. (a) Draw a Venn diagram and work from the inside out. The 10 is given as $|S \cap B \cap J|$. The 17 is calculated from $27 = |J \cap S|$. Etc. Answer = 15.

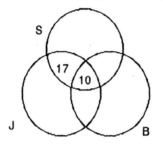

    (b) 71.

4. (a) $\binom{28}{7} = 1,184,040$.   (b) $\binom{12}{3} \cdot \binom{16}{4} = 400,400$.
    (c) $\binom{12}{7} + \binom{16}{7} = 792 + 11,440 = 12,232$.

5. (a) 126.   (b) 105.

6. $\binom{8}{4} \cdot \binom{6}{4} = 1050$.

7. (a) 0.   (b) 840.   (c) 2401.

8. (a) $\binom{9}{4} = 126$.   (b) $5! = 120$.   (c) $5^3 \cdot 9^2 = 10,125$.

9. (a) This is the same as the number of ways of drawing ten cards so that the *first* one is not a repetition. Hence $52 \cdot (51)^9$.
    (b) $52^{10} - 52(51)^9$.

10. (a) 5^k.   (b) $5 \cdot 4 \cdot 3 = 60$.   (c) $4 \cdot 4^3 = 256$.   (d) $5^4 - 4^4 = 369$.

11. (a) $13 \cdot \binom{4}{4} \cdot \binom{48}{1} = 624$.   (b) 5108.

     (c) $13 \cdot \binom{4}{3} \cdot \binom{12}{2} \cdot 4 \cdot 4 = 54{,}912$.   (d) 1,098,240.

12. (a) $2 \cdot 5! = 240$. Treat ab as a single letter and then do the same for ba.

     (b) $6! - 2 \cdot 5! = 480$.

     (c) $2 \cdot 5! - 2 \cdot 4! = 192$. The arrangements in part (a) with bac or cab do not qualify.

13. (a) It is the $n \times n$ matrix with 0's on the diagonal and 1's elsewhere.

     (b) $(n^2 - n)/2$.

14. (a) $n!$ (b) $5 \cdot 4 \cdot 3 \cdot 2 \cdot 1 + 5 \cdot 4 \cdot 3 \cdot 2 + 5 \cdot 4 \cdot 3 + 5 \cdot 4 = 320$.

15. (a) $n \cdot (n-1)^3$. If the vertex sequence is written $v_1 v_2 v_3 v_4$, there are n choices for v_1 and $n - 1$ choices for each of v_2, v_3 and v_4.

     (b) $n(n-1)(n-2)(n-3)$.

     (c) $n(n-1)(n-2)(n-2)$. v_1, v_2, v_3 must be distinct but v_4 can be v_1.

Section 5.2

   1. (a) $\dfrac{8}{25} = .32$.   (b) .20.   (c) $\dfrac{9}{25} = .36$.

   2. $\dfrac{5}{26}$.

   3. (a) $\dfrac{5 \cdot 4 \cdot 3 \cdot 2}{5^4} = .192$.   (b) $\dfrac{3^4}{5^4} = .1296$.   (c) $\dfrac{2}{5} = .40$.

   4. (a) 0.   (b) $\dfrac{2^5}{3^5} \approx .132$.   (c) $\dfrac{1 \cdot 3^4}{3^5} = \dfrac{1}{3}$.

   5. Note that $\binom{7}{3} = 35$ is the number of ways to select three balls from the urn.

     (a) $\dfrac{1}{35}$.   (b) $\dfrac{4}{35}$.   (c) $\dfrac{3 \cdot \binom{4}{2}}{35} = \dfrac{18}{35}$.   (d) $\dfrac{\binom{3}{2} \cdot 4}{35} = \dfrac{12}{35}$.   (e) 1.

   6. (a) $\dfrac{3}{10}$.   (b) $\dfrac{1}{10}$.   (c) $\dfrac{6}{10}$.

7. (a) .2.   (b) .9.   (c) .6.

8. $P(A \cap B) = P(A) + P(B) - P(A \cup B) \geq P(A) + P(B) - 1 = .3.$

9. Here $N = 2,598,960.$
    (a) $624/N \approx .000240.$ (b) $54,912/N \approx .0211.$ (c) $10,200/N \approx .00392.$
    (d) $123,552/N \approx .0475.$ (e) $1,098,240/N \approx .423.$

10. (a) The number of such hands is $4 + 36 + 624 + 3744 + 5108 + 10,200 + 54,912 + 123,552 = 198,180,$ so probability is $198,180/N \approx .0763$ where $N = 2,598,960.$
    (b) 4/13 of the 1,098,240 pairs found in Exercise 11(d), § 5.1, are pairs of Jacks, Queens, Kings or Aces. Adding this to the sum in part (a) yields 536,100. The answer is $\approx .206.$

11. (a) $\frac{1}{2}.$ (b) $P(\{(k,\ell) : k < \ell\}) = \frac{15}{36}.$ (c) $\frac{3}{36} = \frac{1}{12}.$

12. (a) $P(\{(k,\ell) : \max\{k, \ell\} = 4\}) = \frac{7}{36}.$
    (b) $P(\{(k,\ell) : \min\{k, \ell\} = 4\}) = \frac{5}{36}.$
    (c) $P(\{(k,\ell) : k \cdot \ell = 4\}) = \frac{3}{36}.$

13. We have $P(E_1 \cup E_2 \cup E_3) = P(E_1 \cup E_2) + P(E_3) - P((E_1 \cup E_2) \cap E_3).$ Now $P(E_1 \cup E_2) = P(E_1) + P(E_2) - P(E_1 \cap E_2)$ and $P((E_1 \cup E_2) \cap E_3) = P((E_1 \cap E_3) \cup (E_2 \cap E_3) = P(E_1 \cap E_3) + P(E_2 \cap E_3) - P(E_1 \cap E_3 \cap E_2 \cap E_3).$ Substitute. Alternatively, use $E_1 \cup E_2 \cup E_3 = (E_1 \backslash E_2) \cup (E_2 \backslash E_3) \cup (E_3 \backslash E_1) \cup (E_1 \cap E_2 \cap E_3).$

14. F is a disjoint union of E and $F \backslash E$ and so $P(F) = P(E) + P(F \backslash E) \geq P(E).$

15. (a) $\frac{1}{64}.$ (b) $\frac{6}{64}.$ (c) $\frac{15}{64}.$ (d) $\frac{20}{64}.$ (e) $\frac{22}{64}.$

16. $1 - P(\text{tossed} \leq 3 \text{ times}) = 1 - \{\frac{1}{2} + \frac{1}{4} + \frac{1}{8}\} = \frac{1}{8}.$

17. Let Ω_n be all n-tuples of H's and T's and E_n all n-tuples in Ω_n with an even number of H's. It suffices to show that $|E_n| = 2^{n-1},$ which can be done by induction. Assume true for n. Every n-tuple in E_{n+1} has the form (ω, H) where $\omega \in \Omega_n \backslash E_n$ or (ω, T)

where $\omega \in E_n$. There are 2^{n-1} n-tuples of each type, so E_{n+1} has 2^n elements. This problem is easier using conditional probabilities [§ 9.1].

18. With suggestive notation, $1 - P(W_1 \cup W_2) = 1 - [.5 + .4 - .3] = .4.$

19. (a) If Ω is the set of all 4-element subsets of S, the outcomes are equally likely. If E_2 is the event "exactly 2 are even," then $|E_2| = \binom{4}{2} \cdot \binom{4}{2} = 36$. Since $|\Omega| = \binom{8}{4} = 70$,

   $P(E_2) = \frac{36}{70} \approx .514.$

   (b) $\frac{1}{70}$.   (c) $\frac{16}{70}$.   (d) $\frac{16}{70}$.   (e) $\frac{1}{70}$.

20. (a) $\frac{1}{2^3}[3 + 1] = \frac{1}{2}$.   (b) $\frac{1}{2^6}[15 + 6 + 1] \approx .344.$

   (c) $\frac{1}{2^9}[84 + 36 + 9 + 1] \approx .254.$

21. (a) The sample space Ω is all triples (k,ℓ,m) where $k,\ell,m \in \{1, 2, 3\}$, so $|\Omega| = 3^3$. The triples where $k,\ell,m \in \{2, 3\}$ correspond to no selection of 1. So P(1 not selected) $= \frac{2^3}{3^3}$ and answer $= 1 - \frac{8}{27} \approx .704.$

   (b) $1 - \frac{3^4}{4^4} \approx .684.$ (c) $1 - (\frac{n-1}{n})^n.$

   (d) $1 - (.999999)^{1,000,000} \approx .632120.$ This is essentially $1 - \frac{1}{e} \approx .632121$, since

   $\lim_{n \to \infty} (\frac{n}{n-1})^n = e.$

Section 5.3

1. 125.

2. (a) $900 - 7 \cdot 8 \cdot 8 = 452.$ Note that $7 \cdot 8 \cdot 8 =$ number of numbers in S that use neither 3 nor 7.

   (b) Note that $900 - 8 \cdot 9 \cdot 9 = 252$ numbers have at least one 3. Similarly, 252 numbers have at least one 7. So $252 + 252 - 452 = 52$ have a 3 and a 7.

   Alternatively, let S_3 be the numbers in S with no digit that is 3. Similarly for S_7. We seek $900 - |S_3 \cup S_7|$, but

   $|S_3 \cup S_7| = |S_3| + |S_7| - |S_3 \cap S_7| = 8 \cdot 9 \cdot 9 + 8 \cdot 9 \cdot 9 - 7 \cdot 8 \cdot 8 = 848.$

3. There are 466 such numbers in the set, so the probability is .466. Remember that $D_4 \cap D_6 = D_{12}$ not D_{24}. Hence $|D_4| + |D_5| + |D_6| - |D_4 \cap D_5| - |D_4 \cap D_6| - |D_5 \cap D_6| + |D_4 \cap D_5 \cap D_6| = 250 + 200 + 166 - 50 - 83 - 33 + 16 = 466$.

4. (a) $\binom{7 + 3 - 1}{3 - 1} = 36$.

(b) $\binom{4 + 3 - 1}{3 - 1} = 15$. Put $1000 into each fund first, and then invest the remaining $4000.

5. (a) .142. (b) .09.

(c) .78 since $1000 - |D_7 \cup D_{11}| = 1000 - (142 + 90 - 12) = 780$.

(d) .208 since $|D_7 \oplus D_{11}| = |D_7| + |D_{11}| - 2 \cdot |D_7 \cap D_{11}| = 142 + 90 - 2 \cdot 12 = 208$.

6. The formula adds $|\bigcap_{i \in I} A_i|$ if $|I| + 1$ is even, i.e., if I has an odd number of elements, and subtracts $|\bigcap_{i \in I} A_i|$ if $|I|$ is even. Thus it produces the result of the Inclusion-Exclusion Principle.

The formula can be proved by induction on n, using the observation that
$$\mathcal{P}_+(n + 1) = \mathcal{P}_+(n) \cup \{I \cup \{n + 1\} : I \in \mathcal{P}_+(n)\} \cup \{n + 1\}.$$

7. (a) $\binom{12 + 4 - 1}{4 - 1} = 455$.

(b) $\binom{4 + 4 - 1}{4 - 1} = 35$. Put two letters in each box first, and then distribute the remaining four letters.

8. 286, as in Example 6.

9. (a) $x^4 + 8x^3y + 24x^2y^2 + 32xy^3 + 16y^4$.

(b) $x^6 - 6x^5y + 15x^4y^2 - 20x^3y^3 + 15x^2y^4 - 6xy^5 + y^6$.

(c) $81x^4 + 108x^3 + 54x^2 + 12x + 1$.

(d) $x^5 + 10x^4 + 40x^3 + 80x^2 + 80x + 32$.

10. Let $a = -1$ and $b = 1$ in the Binomial Theorem.

11. (a) $\binom{n}{r} = \dfrac{n!}{(n-r)! \cdot r!} = \dfrac{n!}{r! \cdot (n-r)!} = \dfrac{n!}{(n - (n-r))! \cdot (n-r)!} = \binom{n}{n - r}$.

(b) Counting all the r-element subsets of an n-element set is the same as counting all their (n - r)-element complements.

12. (a) Basis: $(a + b)^0 = 1 = \binom{0}{0} \cdot 1 \cdot 1 = \sum\limits_{r=0}^{0} \binom{0}{r} a^r b^{-r}$, since $\binom{0}{0} = \dfrac{0!}{0! \cdot 0!} = \dfrac{1}{1 \cdot 1}$.

Inductive step:

$$(a + b)^n (a + b) = \sum_{r=0}^{n} \binom{n}{r} a^r b^{n-r} (a + b)$$

$$= \sum_{r=0}^{n} \binom{n}{r} a^{r+1} b^{n-r} + \sum_{r=0}^{n} \binom{n}{r} a^r b^{n-r+1}$$

$$= \sum_{s=1}^{n+1} \binom{n}{s-1} a^s b^{n-(s-1)} + \sum_{r=0}^{n} \binom{n}{r} a^r b^{n-r+1} \qquad \text{[using } s = r + 1 \text{ in the first sum]}$$

$$= \binom{n}{n+1-1} a^{n+1} b^{n-n} + \sum_{s=1}^{n} \binom{n}{s-1} a^s b^{n+1-s} + \sum_{r=1}^{n} \binom{n}{r} a^r b^{n+1-r} + \binom{n}{0} a^0 b^{n+1}$$

$$= a^{n+1} b^0 + \sum_{t=1}^{n} \left\{ \binom{n}{t-1} + \binom{n}{t} \right\} a^t b^{n+1-t} + a^0 b^{n+1}$$

$$= a^{n+1} b^0 + \sum_{t=1}^{n} \binom{n+1}{t} a^t b^{n+1-t} + a^0 b^{n+1} \qquad \text{[using the relation proved following the statement of the Binomial Theorem]}$$

$$= \sum_{t=0}^{n+1} \binom{n+1}{t} a^t b^{n+1-t}.$$

(b) $\binom{n}{r-1} + \binom{n}{r} = \dfrac{n!}{(r-1)! \, (n-r+1)!} + \dfrac{n!}{r! \, (n-r)!}$

$$= \dfrac{n!}{r! \, (n-r+1)!} \cdot [r + (n-r+1)] = \dfrac{(n+1)!}{r! \, (n-r+1)!} = \binom{n+1}{r}.$$

13. (b) There are $\binom{n}{r}$ subsets of size r for each r, so there are $\sum\limits_{r=0}^{n} \binom{n}{r}$ subsets in all.

(c) If true for n, then

$$\sum_{r=0}^{n+1} \binom{n+1}{r} = 1 + \sum_{r=1}^{n} \binom{n+1}{r} + 1 = 1 + \sum_{r=1}^{n} \binom{n}{r-1} + \sum_{r=1}^{n} \binom{n}{r} + 1$$

$$= \sum_{r=1}^{n+1} \binom{n}{r-1} + \sum_{r=0}^{n} \binom{n}{r} = 2 \sum_{r=0}^{n} \binom{n}{r} = 2 \cdot 2^n = 2^{n+1}.$$

14. Set $a = 2$ and $b = 1$ in the Binomial Theorem.

15. (a) $\sum\limits_{k=3}^{5} \binom{k}{3} = \binom{3}{3} + \binom{4}{3} + \binom{5}{3} = 15 = \binom{6}{4}$.

(b) Let $p(n)$ be " $\sum\limits_{k=m}^{n} \binom{k}{m} = \binom{n+1}{m+1}$ " for $n \geq m$. Check that $p(m)$ is true. Assume that $p(n)$ is true for some $n \geq m$. Then

$$\sum_{k=m}^{n+1} \binom{k}{m} = \left[\sum_{k=m}^{n} \binom{k}{m} \right] + \binom{n+1}{m}$$

$$= \binom{n+1}{m+1} + \binom{n+1}{m} \qquad \text{[by the inductive assumption]}$$

76

$$= \binom{n+2}{m+1} \qquad \text{[by Exercise 12(b)].}$$

(c) The set $\mathscr{A}$ of $(m+1)$-element subsets of $\{1, 2, \ldots, n+1\}$ is the disjoint union $\bigcup_{k=m}^{n} \mathscr{A}_k$, where $\mathscr{A}_k$ is the collection of $(m+1)$-element subsets whose largest element is $k+1$. A set in $\mathscr{A}_k$ is an m-element subset of $\{1, 2, \ldots, k\}$ with $k+1$ added to it, so

$$|\mathscr{A}| = \sum_{k=m}^{n} \binom{k}{m}.$$

16. (a)

(b) 0 1 0 1 0 1 0 0, etc.

17. (a) Put 8 in one of the 3 boxes, then distribute the remaining 6 in three boxes. Answer is $3 \cdot \binom{8}{2} = 84$ ways.

(b) $\binom{16}{2} - 3 \cdot \binom{8}{2} = 36.$

(c) Since $1 + 9 + 9 < 20$, each digit is at least 2. Then $(d_1 - 2) + (d_2 - 2) + (d_3 - 2) = 20 - 6 = 14$ with $2 \le d_i \le 9$ for $i = 1, 2, 3$. By part (b) there are 36 numbers.

Another way to get this number is to hunt for a pattern:

299; 398,389; 497,488,479; 596,587,578,569; ... . Looks like

$1 + 2 + 3 + \cdots + 8 = \binom{9}{2} = 36.$

Section 5.4

1. (a) $\dfrac{15!}{3! \, 4! \, 5! \, 3!}.$ (b) $\binom{15}{3}\binom{15}{4}\binom{15}{5}.$

2. $\binom{n}{k} \cdot \binom{n-k}{r} = \dfrac{n!}{k! \cdot (n-k)!} \cdot \dfrac{(n-k)!}{r! \cdot (n-k-r)!}$, so both numbers are the same in each case.

3. (a) As in Example 4, count ordered partitions $\{A, B, C, D\}$ where $|A| = 5$, $|B| = 3$, $|C| = 2$ and $|D| = 3$. Answer $= \dfrac{13!}{5! \cdot 3! \cdot 2! \cdot 3!} = 720{,}720.$

(b) Count ordered partitions where $|A| = 4$ and $|B| = |C| = |D| = 3$, but note that partitions $\{A, B, C, D\}$ and $\{A, C, B, D\}$ are equivalent. Answer $= \dfrac{1}{2} \cdot \dfrac{13!}{4! \cdot 3! \cdot 3! \cdot 3!} = 600{,}600.$

(c) Count ordered partitions where $|A| = |B| = |C| = 3$ and $|D| = 4$, but note that permutations of A, B and C give equivalent sets of committees. Answer $= \dfrac{1}{6} \cdot \dfrac{13!}{3! \cdot 3! \cdot 3! \cdot 4!}$

$= 200{,}200.$

4. $9!/(3!\cdot2!\cdot4!) = 1260.$

5. (a) $3^{10} = 59,049.$ (b) $\binom{10}{5} = 252.$

   (c) $\binom{10}{3} = 120.$ (d) $\binom{10}{3}\cdot2^7 = 15,360.$

   (e) $\dfrac{10!}{3!\cdot4!\cdot3!} = 4200.$

   (f) $3^{10} - 3\cdot2^{10} + 3\cdot1 - 1\cdot0 = 55,980,$ using the Inclusion- Exclusion Principle on the sets of sequences with no 0's, no 1's and no 2's.

6. (a) $7! = 5040.$ (b) $\dfrac{10!}{2!\cdot2!} = 907,200.$

   (c) $\dfrac{11!}{4!\cdot4!\cdot2!} = 34,650.$ (d) $\dfrac{4!}{2!} = 12.$

7. (a) $625.$ (b) $5^4 - 5\cdot4\cdot3\cdot2 = 505.$ (c) $5^3\cdot2 = 250.$

   (d) $(3/5)\cdot505 = 303.$ Or $3\cdot5^3 - 3\cdot4\cdot3\cdot2 = 303.$

8. These are like Example 5(c). For instance, there are $3!\cdot6!/(4!\cdot2!) = 90$ words of type 4-2-0, because there are $3!$ ways to choose the letters with 4, 2 and 0 occurrences, but there are $3\cdot6!/(3!\cdot3!) = 60$ words of type 3-3-0, because there are 3 ways to choose which letter does not occur.

9. There are $\frac{1}{2}\binom{2n}{n}$ unordered such partitions and $\binom{2n}{n}$ ordered partitions.

10. (a) The five vowels partition the set of consonants into six strings of consecutive consonants. Since there are 21 consonants, at least one string must have 4 or more consonants.
    (b) One such list is b c d f a g h j k e ℓ m n p i q r s t o v w x y u z.
    (c) Now the five vowels partition the set of consonants into *five* strings of consecutive consonants and so at least one string must have 5 or more consonants.

11. (a) $10\cdot\binom{8}{2}\cdot\binom{5}{2}\cdot\binom{2}{2} = 2800.$ Just choose a third member of their team and then choose 3 teams from the remaining 9 contestants.

    (b) $10/\binom{11}{2} = 2/11 \approx .18.$

12. $\dfrac{16!}{2^8\cdot8!} = 15\cdot13\cdot11\cdot9\cdot7\cdot5\cdot3\cdot1 = 2,027,025.$

13. (a) Think of putting 9 objects in 4 boxes, starting with 1 object in the first box.
$\binom{8+4-1}{4-1} = 165$.

(b) Now start with 1 object in each box; $\binom{5+4-1}{4-1} = 56$.

14. The function $\chi_A + \chi_B$ has value 2 on the set $A \cap B$ and value 1 on the set $A \oplus B$. Similarly for $\chi_C + \chi_D$, so $A \cap B = C \cap D$ and $A \oplus B = C \oplus D$. Hence $|A| + |B| = |A \oplus B| + 2|A \cap B| = |C \oplus D| + 2|C \cap D| = |C| + |D|$.

15. Fifteen. Just count. We know no clever trick, other than breaking them up into types 4, 3-1, 2-2, 2-1-1, and 1-1-1-1. There are 1, 4, 3, 6 and 1 partitions of these respective types.

This solves the problem because there is a one-to-one correspondence between equivalence relations and partitions; see Theorem 1 of § 3.5.

16. 1/1680. This is N O N S E N S E revisited.

Section 5.5

1. (a) Apply the Pigeon-Hole Principle to the partition $\{A_0, A_1, A_2\}$ of the set S of four integers where $A_i = \{n \in S : n \equiv i \pmod 3\}$. Alternatively, apply the second version of the Pigeon-Hole Principle to the function MOD 3: $S \to \mathbb{Z}(3)$.

(b) Apply the Pigeon-Hole Principle to the function $f: \{1, 2, \ldots, p + 1\} \to \mathbb{Z}(p)$ defined by $f(m) = a_m$ MOD p.

2. (a) By the Pigeon-Hole Principle, there are at least $50/4 = 12.5$ marbles of some color, so there are at least 13 of that color.

(b) Use the Pigeon-Hole Principle with the remaining 42 marbles and 3 colors.

3. Here $|S| = 73$ and $73/8 > 9$, so some box has more than 9 marbles.

4. (a) For each ordered pair (a,b) in B let $f(a,b) = a + b$. There are only 11 possible values of $f(a,b)$, namely $2, 3, \ldots, 12$. Apply the second version of the Pigeon-Hole Principle.

(b) At most 11 times; see part (a).

5. For each 4-element subset B of A, let $f(B)$ be the sum of the numbers in B. Then $f(B) \geq 1 + 2 + 3 + 4 = 10$, and $f(B) \leq 50 + 49 + 48 + 47 = 194$. There are thus only 185

possible values of $f(B)$. Since A has $\binom{10}{4} = 210$ subsets B, at least two of them must have the same sum $f(B)$ by the second version of the Pigeon-Hole Principle.

6. For each nonempty subset B of S let $f(B)$ be the element of $\mathbb{Z}(6)$ that is congruent to the sum of the numbers in B. Since S has $2^3 - 1 = 7$ nonempty subsets and $f(B)$ takes at most six values, at least two nonempty subsets B of S have the same sum modulo 6, by the second version of the Pigeon-Hole Principle.

7. For each 2-element subset T of A, let $f(T)$ be the sum of the two elements. Then f maps the 300 2-element subsets of A into $\{3, 4, 5, \ldots, 299\}$.

8. (a) There is no such sequence. This shows that the number $n^2 + 1$ in Example 4 cannot be replaced by n^2; here $n = 4$.
   (b) Example 4 guarantees that there is at least one example. One such is $17, 13, 9, 5, 1$. All contain 17.
   (c) $2, 3, 7, 9, 15$ is one example. $17, 16, 13, 11, 9, 4, 1$ is an example of length 7.

9. The repeating blocks are various permutations of 142857.

10. If $1 \le k \le n$ and if n nonnegative numbers have sum m, then there must be k of them with sum at least km/n. One way to show this is the first approach in Example 6(b).
    An alternative is a proof by induction on k, as in the third approach to that example. Assume that $a_1 \ge a_2 \ge \cdots \ge a_n$ and let $s = a_1 + \cdots + a_k$ for the inductive step. Then
    $$m - s = a_{k+1} + \cdots + a_n \le (n - k)a_{k+1}$$
    and hence $a_{k+1} \ge (m - s)/(n - k)$. The inductive assumption is that $s \ge km/n$. Thus
    $$a_1 + \cdots + a_k + a_{k+1} = s + a_{k+1} \ge s + (m - s)/(n - k) = s[1 - 1/(n-k)] + m/(n-k)$$
    $$\ge (km/n)[1 - 1/(n-k)] + m/(n-k) \qquad [\text{since } 1 > \frac{1}{n-k} \text{ for } k < n]$$
    $$= (km/n) + (m - km/n)/(n-k) = (km/n) + m/n = (k + 1)m/n.$$

11. (a) Look at the six blocks
    $$(n_1, n_2, n_3, n_4), \quad (n_5, n_6, n_7, n_8), \quad \ldots, \quad (n_{21}, n_{22}, n_{23}, n_{24}).$$
    (b) Use Example 2(b) of § 4.2.
    (c) Look at the 8 blocks $(n_1, n_2, n_3), (n_4, n_5, n_6), \ldots, (n_{22}, n_{23}, n_{24})$. By part (b), at least one block has sum at least $300/8 = 37.5$.
    (d) Look at the five blocks
    $$(n_1, \ldots, n_5), \quad (n_6, \ldots, n_{10}), \quad (n_{11}, \ldots, n_{15}), \quad (n_{16}, \ldots, n_{20}), \quad (n_{21}, \ldots, n_{24}).$$

Note that the last block has only 4 members. If some block has sum greater than 300/5 we are done. Otherwise, all have sum exactly 60, and adjoining an element to the last block gives sum at least 61. •

12. (a) The average of all such sums is $18.5 \cdot 4 = 74$.

(b) As in Example 7(b), group the 36 sectors into 9 disjoint consecutive blocks of 4. If all blocks sum to exactly 74, shift clockwise one sector and get new sums.

(c) The average of all such sums is $18.5 \cdot 5 = 92.5$.

(d) Ignore the sector numbered 1 and group the other 35 sectors into 7 disjoint consecutive blocks of 5. Since $\frac{666 - 1}{7} = 95$, some block has sum at least 95.

13. If $0 \in \text{Im}(f)$, then $n_1, n_1 + n_2$ or $n_1 + n_2 + n_3$ is divisible by 3. Otherwise f is not one-to-one and there are three cases:
If $n_1 \equiv n_1 + n_2 \pmod 3$, then $n_2 \equiv 0 \pmod 3$ and n_2 is divisible by 3.
Similarly, if $n_1 \equiv n_1 + n_2 + n_3 \pmod 3$, then $n_2 + n_3$ is divisible by 3.
And if $n_1 + n_2 \equiv n_1 + n_2 + n_3 \pmod 3$, then n_3 is divisible by 3.

14. There are $\binom{15}{5} = 3003$ such committees in all.

(a) Since $\binom{6}{2} \cdot \binom{9}{3} = 1260$, the probability is $\frac{1260}{3003} \approx .420$.

(b) Since $\binom{15}{5} - \binom{9}{5} - \binom{6}{5} = 2871$, the probability is $\frac{2871}{3003} \approx .956$.

(c) Since $\binom{9}{5} + \binom{6}{5} = 132$, the probability is $\frac{132}{3003} \approx .044$.

15. (a) $8^6 = 262,144$.

(b) Let A be the set of numbers with no 3's and B be the set with no 5's. We want $|A^c \cap B^c| = |(A \cup B)^c|$. Since $|A| = |B| = 7^6$ and $|A \cap B| = 6^6$, $|A \cup B| = 2 \cdot 7^6 - 6^6$ and the answer is $8^6 - (2 \cdot 7^6 - 6^6) = 73,502$.

(c) $8!/2! = 20,160$. (d) $6!/(1! \cdot 2! \cdot 3!) = 60$.

16. $(4 + 1) \cdot (1 + 1) \cdot (3 + 1) = 40$.

17. (a) We show that S must contain both members of some pair $(2k-1, 2k)$. Partition $\{1, 2, \ldots, 2n\}$ into the n subsets $\{1, 2\}, \{3, 4\}, \ldots, \{2n-1, 2n\}$. Since $|S| = n+1$, some subset $\{2k-1, 2k\}$ contains 2 members of S by the Pigeon-Hole Principle. The numbers $2k-1$ and $2k$ are relatively prime.

(b) For each $m \in S$, write $m = 2^k \cdot j$ where j is odd and let $f(m) = j$. See Exercise 21, § 4.1. Then $f: S \to \{1, 3, 5, \ldots, 2n - 1\}$. By the second version of the Pigeon-Hole Principle, there are distinct m and m' in S with $m = 2^k \cdot n$ and $m' = 2^{k'} \cdot n$ for some odd n. Say $k > k'$. Then $m = 2^{k-k'} \cdot m'$, so m' divides m.

(c) Let $S = \{2, 4, 6, \ldots, 2n\}$.

(d) Let $S = \{n + 1, n + 2, \ldots, 2n\}$.

18. (a) Let $B = \{p - n : n \in A\}$. Then $|A \cap B| = |A| + |B| - |A \cup B| \geq |A| + |B| - (p + 1) > 2[(p/2) + 1] - (p + 1) = 1$. Hence $|A \cap B| \geq 2$ and there are distinct $m, n \in A$ such that $p - m, p - n \in A$. If $m \neq p - m$, use m and $p - m$. If $m = p - m$, use n and $p - n$.

(b) and (c)   One example is $\{0, 1, 2, 3\}$.

19. (a)   By the remark at the end of the proof of the Generalized Pigeon-Hole Principle, the average size is $2 \cdot 21/7 = 6$.

(b)   One way is to number the students 0 to 20 and group them as follows:
$\{0, 1, 2, 3, 4, 5\}$, $\{6, 7, 8, 9, 10, 11\}$, $\{12, 13, 14, 15, 16, 17\}$, $\{18, 19, 20, 0, 1, 2\}$,
$\{3, 4, 5, 6, 7, 8\}$, $\{9, 10, 11, 12, 13, 14\}$, $\{15, 16, 17, 18, 19, 20\}$.

CHAPTER 6

This chapter gives the core material on graphs and trees. Chapters 7 and 8 go into more detail and present the algorithms. Some of the terminology has already been introduced in § 3.2.

Section 6.1 treats undirected graphs. Cycles and acyclicity are key ideas. One way to illustrate the truth of Theorem 1 is to draw a snarly path and cut off side trips with an eraser. It's hard to illustrate the proof of Theorem 2, but a picture can make its contrapositive seem plausible. In earlier editions we made some fuss over graph isomorphism. Our approach now is that isomorphism is merely a precise way of saying "look just alike". The idea of the degree of a vertex is of course basic, independent of any discussion of isomorphism.

Section 6.2 contains a plausible but phony proof of Euler's theorem just after Example 2. It's a good idea to go slowly through a counterexample to the "proof". Then see how the real proof handles the example and watch FLEURY'S algorithm construct a path. This section also introduces connectedness. Draw a picture to illustrate the ideas in Theorem 3. Emphasize: (1) the theorem is natural and (2) it gives a link between cycles and connectedness, so a test for connectedness gives a test for cycles.

Trees have come up earlier, but § 6.3 gives the first careful discussion of them. Theorems 2 and 3 will be needed later. Most of the trees in computer science are rooted, and hence directed in a coherent way. Section 6.4 introduces basic terminology. Other terms that we do not use but that students may have heard in this connection are **fan-in** for indegree and **fan-out** for outdegree. Essentially every rooted tree that we see is ordered in some natural way, so we can think of order as a way of organizing a typical rooted tree.

Section 6.5 is less intuitive than § 6.2. Contrast the statements of Theorems 1 and 2. Ask students how hard it would be to verify the hypotheses of Theorem 3. [It takes $O(n^2)$ time.] Point out that its proof is nonconstructive.

Gray codes are an important application and our account just scratches the surface. Try to get class members who are familiar with Gray codes from computer science to say a few words about them. [For instance, Gray code hardware is readily available.] Theorem 4 need not be stressed.

Section 6.6 gives three algorithms for finding spanning trees and forests. The first one, which builds one tree at a time, takes no account of weights for edges. It's actually a stripped-down version of PRIM'S algorithm. We have included it because when weights don't matter it's faster than PRIM'S and it shows the idea of PRIM'S algorithm free of distractions. The other two algorithms we give, KRUSKAL'S and PRIM'S, construct minimum weight spanning forests. Exercise 11 provides good motivation for considering them. KRUSKAL'S algorithm requires an initial sorting of the edges by weight. It builds a graph that is always acyclic and light in weight and eventually is as connected as possible. We recommend going through the proof of Theorem 2 with an example. PRIM'S algorithm is easy to follow, since it works like the unweighted algo-rithm TREE, but its verification is tricky. As with KRUSKAL'S algorithm, a loop invariant is the key. A classroom example can illustrate the invariant.

This section contains detailed time-complexity analyses of all three algorithms. Now is the time to decide how much emphasis you want to place on such analyses. The ones in this section are easy, but if you omit discussing them now, students will find it hard to get interested in or to understand the analyses later on in Chapters 7 and 8.

Section 6.1

1. (a) s t v or s u v; length 2.
   (c) u v w y; length 3.
   (b) s t v w x z and 3 others; length 5.
   (d) v w; length 1.

2. (a) s u t v is one.
   (c) u s t u v w y x z y.
   (b) s t u v w x y z is one.
   (d) v u s t v w is one.
   There are no longest paths connecting vertices. For example, one can go from s to v as follows: s u v u v u v ··· u v.

3. (a) is true and (b) is false. To see the failure of (b), consider the graph in Figure 6(a).

4. (a) and (b) are both true.

5. (a) and (b) are now both true.

6. (a) 12, 6; 14, 7; 16, 8; 20, 10. In each case, the sum of the degrees is twice the number of edges.
   (b) No. By Theorem 3 the degree sum is even, whereas the sum would be odd if there were an odd number of vertices of odd degree.

7. See Figure 1(c). u and w are vertices of a cycle, and so are w and x. But no cycle uses both u and x.

8. If a cycle contains a loop at v, its entire vertex sequence must be v v, since otherwise v would be repeated in the middle of the cycle.

9. (a)

e	a	b	c	d	e	f	g	h	k
$\gamma(e)$	{w, x}	{x, u}	{t, u}	{t, v}	{u, v}	{u, y}	{v, z}	{x, y}	{y, z}

(b) a = {w, x}, b = {x, u}, c = {u, t}, d = {t, v}, e = {u, v}, f = {u, y }, g = {v, z}, h = {x, y}, k = {y, z}.

10.

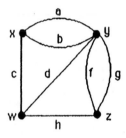

11. (a) e b h k g and its reversal g k h b e.

(b) b c d g k h and its reversal.

(c) e c d, b h f and their reversals.

12. (a)

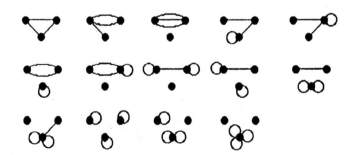

85

(b)

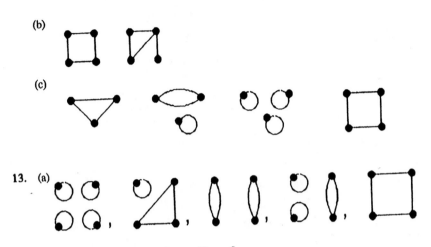

(c)

13. (a)

(b) The only such graph is K_4; see Figure 5.

(c) There are none by Theorem 3, since 5·3 is not even.

14. (a) 1   (b) (0,1,2,3,1).

(c) There is just the identity isomorphism. The vertices w and u must correspond to themselves, since they are the only vertices of degrees 1 and 4. Then x is the only vertex joined to both w and u, t is the only vertex of degree 2 joined to u, z is the only other vertex of degree 2, etc.

(d) There is just one isomorphism. [Otherwise going over to H with one and back to G with the inverse of another would give a non- identity isomorphism of G onto itself, contrary to part (c).]

15. (a), (c) and (d) are regular, but (b) is not. (a) and (c) have cycles of length 3, but (d) does not. Or count edges. (a) and (c) are isomorphic; the labels show an isomorphism between (a) and (c).

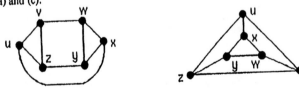

16. The labels show one isomorphism.

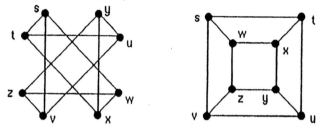

17. (a) $\binom{8}{5} = 56$.

(b) 1 with 1 edge, 6 with 2 edges and $6 \cdot 5 = 30$ with 3 edges, so 37 altogether.

(c) $8 \cdot 7 + 8 \cdot 7 \cdot 6 + 8 \cdot 7 \cdot 6 \cdot 6 = 2{,}408$.

18. (a) By Theorem 3, $42 = 1 \cdot 7 + 2 \cdot 3 + 3 \cdot 7 + 4 \cdot D_4(G)$, so $4 \cdot D_4(G) = 8$, $D_4(G) = 2$ and the total number of vertices is $7 + 3 + 7 + 2 = 19$.

(b) The graph would have $19 + 6 = 25$ vertices.

19. (a)   (b)

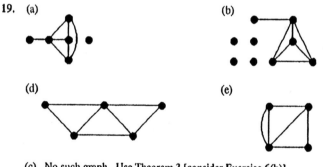

(c) No such graph. Use Theorem 3 [consider Exercise 6(b)].

(f) No such graph. Use Theorem 3.   (g) K_4.   (h) K_5.

20. Consider a cycle in a graph G, with vertex sequence $x_1 x_2 \cdots x_n x_1$. Then $x_1, \ldots, x_n$ are distinct and no edge appears twice. Assign directions to the edges in the cycle so they go from x_1 to x_2 to x_3 to $\ldots$ to x_n to x_1. Since no edge appears twice, this assignment cannot contradict itself. Assign directions arbitrarily to the other edges of G. Then the edges in the cycle form a directed cycle in the resulting digraph.

Conversely, suppose it is possible to assign directions to edges in G so that a path with vertex sequence $x_1 x_2 \cdots x_m x_1$ is a directed cycle. Then $m \geq 1$ and $x_1, \ldots, x_m$ are all different. If $m \geq 3$ then the path is an undirected cycle by Proposition 1. If $m = 2$ the vertex sequence is $x_1 x_2 x_1$, but the edges from x_1 to x_2 and from x_2 to x_1 must be different because their directions are different. In this case the undirected path is simple, so it is a cycle. Finally, if $m = 1$ the path is a loop, so it's a cycle. In all cases the directed cycle is an undirected cycle.

21. Assume no loops or parallel edges. Consider a longest path $v_1 \cdots v_m$ with distinct vertices. There is another edge at v_m. Adjoin it to the path to get a closed path and use Proposition 1.

22. If $n = 1$ the graph has at least one loop, so it contains a cycle. Suppose inductively that the assertion is true for some $n \in \mathbb{P}$ and consider a graph G with $n + 1$ vertices and at least $n + 1$ edges. If every vertex has degree at least 2 the graph contains a cycle by Exercise 21. Otherwise, G has a vertex v of degree 0 or degree 1. In that case, delete v and the edge to it if there is one. What's left has n vertices and at least n edges, so it contains a cycle by the inductive assumption. Hence G itself contains a cycle. The claim follows by induction.

23. Use $|V(G)| = D_0(G) + D_1(G) + D_2(G) + \cdots$ and Theorem 3.

24. (a) The identity mappings on V(G) and E(G) define an isomorphism, so $G \simeq G$ and $\simeq$ is reflexive. If α and β define an isomorphism of G onto H, then $\alpha^{\leftarrow}$ and $\beta^{\leftarrow}$ define an isomorphism of H onto G. Thus $\simeq$ is symmetric.

Finally $\simeq$ is transitive as follows. The inverses of $\alpha_2 \circ \alpha_1$ and $\beta_2 \circ \beta_1$ are $\alpha_1^{\leftarrow} \circ \alpha_2^{\leftarrow}$ and $\beta_1^{\leftarrow} \circ \beta_2^{\leftarrow}$, respectively, so both $\alpha_2 \circ \alpha_1$ and $\beta_2 \circ \beta_1$ are one-to-one correspondences. Say G, H and K are described by γ_1, γ_2 and γ_3. Suppose $e \in E(G)$ and $\gamma_1(e) = \{u, v\}$. Then, since α_1 and β_1 describe an isomorphism, $\gamma_2(\beta_1(e)) = \{\alpha_1(u), \alpha_1(v)\}$. Since α_2 and β_2 describe an isomorphism, whenever $f \in E(H)$ and $\gamma_2(f) = \{w, x\}$ then $\gamma_3(\beta_2(f)) = \{\alpha_2(w), \alpha_2(x)\}$. In particular, $\gamma_3(\beta_2(\beta_1(e))) = \{\alpha_2(\alpha_1(u)), \alpha_2(\alpha_1(v))\}$. This shows that $\gamma_3((\beta_2 \circ \beta_1)(e)) = \{(\alpha_2 \circ \alpha_1)(u), (\alpha_2 \circ \alpha_1)(v)\}$ whenever $\gamma_1(e) = \{u, v\}$, i.e., $\alpha_2 \circ \alpha_1$ and $\beta_2 \circ \beta_1$ describe an isomorphism.

(b) There are 2 classes; see Example 4.

Section 6.2

1. Only Figure 7(b) has an Euler circuit. To find one, do Exercise 2.

2. r s t v s u v w z y v x y u r is the sequence obtained by choosing the earliest letter in the alphabet at each Fleury choice.

3. It won't break down until the second visit to the other vertex of degree 3, namely t.

4. After starting out with w x u y v or w x u y w z any choice disconnects the graph.

5. (a) $v_3 v_1 v_2 v_3 v_6 v_2 v_4 v_6 v_5 v_1 v_4 v_5 v_3 v_4$ is one.
(b) There is no Euler circuit since v_3 and v_4 have odd degree.

6. There is no Euler path or circuit since four vertices have odd degree.

7. No. The edges and corners form a graph with eight vertices, each of degree 3. It has no Euler path. See Figure 5(c) of § 6.5.

8. Here is one possible result, starting at v.

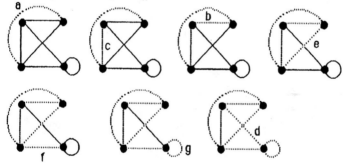

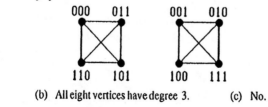

The edge sequence is a c b e f g d.

9. $\{0, 1\}^3$ consists of 3-tuples of 0's and 1's, which we may view as binary strings of length 3. The graph is then

```
  000      011      001      010
   ·────────·        ·────────·
   │ ╲    ╱ │        │ ╲    ╱ │
   │   ╳    │        │   ╳    │
   │ ╱    ╲ │        │ ╱    ╲ │
   ·────────·        ·────────·
  110      101      100      111
```

(a) 2. 　　　　(b) All eight vertices have degree 3. 　　　　(c) No.

10. View $\{0, 1\}^3$ as binary strings of length 3. The graph is then:

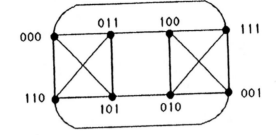

(a) There is one component.

(b) All eight vertices have degree 4.

(c) Yes, by Euler's theorem.

11. (a) Join the odd-degree vertices in pairs, with k new edges. The new graph has an Euler circuit, by Theorem 2. The new edges do not appear next to each other in the circuit and they partition the circuit into k simple paths of G.

(b) One solution is:

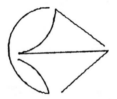

(c) Imitate the proof. That is, add edges $\{v_2, v_3\}$ and $\{v_5, v_6\}$, say, create an Euler circuit, and then remove the two new edges. Then you will obtain $v_3 v_1 v_2 v_7 v_3 v_4 v_5 v_7$ $v_6 v_1 v_5$ and $v_6 v_4 v_2$, say.

12. The ones with n odd.

13. (a) No such walk is possible. Create a graph as follows. Put a vertex in each room and one vertex outside the house. For each door draw an edge joining the vertices for the regions on its two sides. The resulting graph has two vertices of degree 4, three of degree 5 and one of degree 9. Apply the corollary to Theorem 1.

(b) An Euler path starting outside is possible. In this case the graph has four vertices of degree 4, one of degree 5 and one of degree 9.

90

Section 6.3

1.

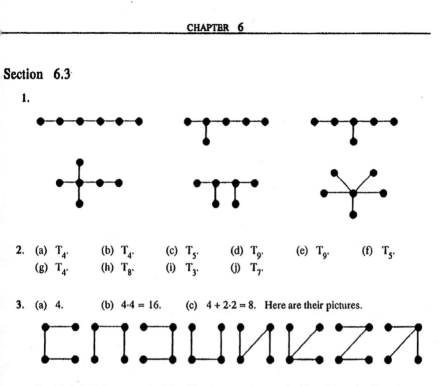

2. (a) T_4.   (b) T_4.   (c) T_5.   (d) T_9.   (e) T_9.   (f) T_5.
(g) T_4.   (h) T_8.   (i) T_3.   (j) T_7.

3. (a) 4.   (b) $4 \cdot 4 = 16$.   (c) $4 + 2 \cdot 2 = 8$.   Here are their pictures.

For (d) and (e) the answer is $8 \cdot 8 = 64$. Any spanning tree in (d) or (e) can be viewed as a pair of spanning trees, one for the upper half and one for the lower half of the graph. Each half of the spanning tree is a spanning tree for the graph in (c). Hence there are $8 \cdot 8 = 64$ such pairs.

4. The pictures

illustrate the three possible isomorphism types.

5. (a) Since $2n - 2$ is the sum of the degrees of the vertices, we must have $2n - 2 = 4 + 4 + 3 + 2 + 1 \cdot (n - 4)$. Solve for n to get $n = 11$.

(b) Here is one example:

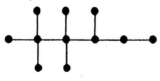

6. (a) If the tree has n vertices, then it must have n - 7 of degree 1. So we must have
$$5 + 5 + 3 + 3 + 3 + 2 + 2 + 1 \cdot (n - 7) = 2n - 2$$
as noted in Exercise 5. Solve for n to get n = 18.

(b) Here is an example:

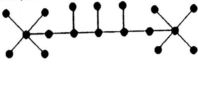

7. (a)

(b) Prove by induction on n. The cases n = 1, 2, 3 are easy. Assume that the result is true for some $n \geq 4$. Suppose that $d_1 + \cdots + d_{n+1} = 2(n + 1) - 2$. At least one d_k is 1, say $d_{n+1} = 1$. At least one d_k exceeds 1, say d_1. Define $d_1^* = d_1 - 1$ and $d_k^* = d_k$ for $2 \leq k \leq n$. Then $d_1^* + \cdots + d_n^* = 2n - 2$ and by the inductive hypothesis there is a tree with n vertices whose vertices have degrees $d_1^*, \ldots, d_n^*$. Attach a leaf to the vertex of degree d_1^* to obtain a tree with n + 1 vertices. The new vertex has degree $1 = d_{n+1}$ and the vertex of degree d_1^* now has degree $d_1^* + 1 = d_1$.

(c) $d_1 = 5$, $d_2 = 3$, $d_3 = 2$, $d_4 = \cdots = d_9 = 1$ and their sum is $16 = 2 \cdot 9 - 2$.

8. There are nine.

9. (a) Suppose the components have $n_1, n_2, \ldots, n_m$ vertices, so that altogether $n_1 + n_2 + \cdots + n_m = n$. The i th component is a tree, so it has $n_i - 1$ edges, by Theorem 3. The total number of edges is $(n_1 - 1) + (n_2 - 1) + \cdots + (n_m - 1) = n - m$.

(b) A spanning forest for the graph has m components, so it has n - m edges, by part (a). Thus the spanning forest is the whole graph.

10. Use Theorem 1 and note that a spanning tree of the graph will have n - 1 edges by Theorem 3.

11. By Lemma 1 to Theorem 3 it must be infinite. Use $\mathbb{Z}$ for the set of vertices; see Figure 3(a), § 13.3.

Section 6.4

1. (a) (b) 1. (c) 3.

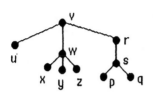

2. Here is one.

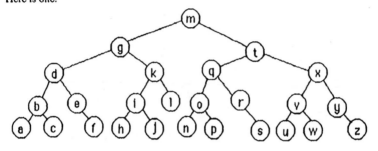

3. (a) Rooted trees in Figures 5(b) and 5(c) have height 2; the one in 5(d) has height 3.

   (b) Only the rooted tree in Figure 5(c) is a regular 3-ary tree.

4. People have two parents, whereas vertices in a rooted tree have at most one parent.

5. (a)

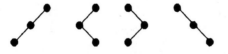

   (b) The fifth and seventh. (c) The seventh.

   (d) 21. There are 4 corresponding to the first type of tree in (a), namely

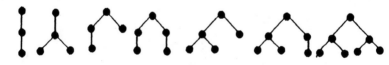

   Similarly, there are 2, 4, 4, 2, 4 and 1 of each of the other types of trees in (a).

6. (a)

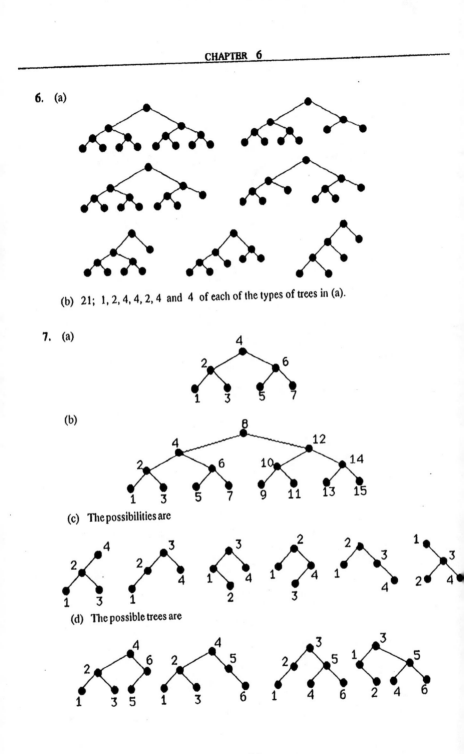

(b) 21; 1, 2, 4, 4, 2, 4 and 4 of each of the types of trees in (a).

7. (a)

(b)

(c) The possibilities are

(d) The possible trees are

8.

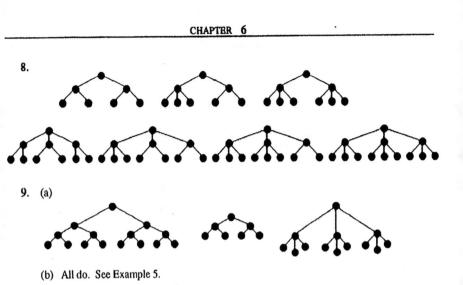

9. (a)

(b) All do. See Example 5.

10. (a) 2^h.   (b) $2^{h+1} - 1$.

11. From Example 5, we have $(m-1)p = m^h - 1$, so $(m-1)p + 1 = m^h = t$.

12. Examples can be found by looking at federal laws, handbooks of tables, parts catalogs, etc.

13. There are 2^k words of length k.

14. Use the last tree in our answer to Exercise 9(a) and label the vertices λ, a, b, c, aa, ab, ac, ba, etc.

15. (a) Move either Lyon's or Ross's records to the Rose node and delete the empty leaf created.
(b) There are three possible good choices, leading to:

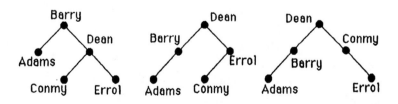

16. (a) The two revisions that only affect the right branch yield:

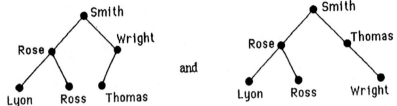

and

(b) The only possibility is a major rearrangement, resulting in:

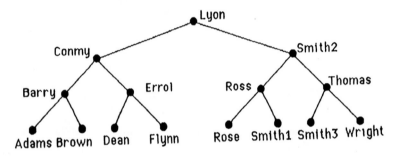

(c) If four clients were added, there would be sixteen clients in all, so the height of the binary search tree would have to be increased.

Section 6.5

1. (a) In the vertex sequence for a Hamilton circuit [if one existed], the vertex w would have to both precede and follow the vertex v. I.e., the vertex w would have to be visited twice.

   (b) In the vertex sequence for a Hamilton path [if one existed], the vertices of degree 1 would have to be at the beginning or end, since otherwise the adjacent vertex would be repeated. But there are three vertices of degree 1 in Figure 1(d).

2. (a) Any path that used every vertex would have to use the central vertex more than once. So no Hamilton circuit exists.

   (b) Note that each edge connects an upper vertex to a lower vertex. If there were a Hamilton circuit, its vertex sequence would alternate between upper and lower vertices and would consist of exactly 8 vertices. The first and eighth vertices would have to be different [one would be an upper one and one would be lower] and yet they would have to be the same to complete the circuit. So no such circuit exists.

(c) If there were a Hamilton circuit, consider its vertex sequence and the vertices just before and after the central vertex v. If these three vertices and the edges connected to them were removed from the graph, the remaining graph would have two components. No vertex in one component could be connected to a vertex in the other component without visiting one of the three removed vertices. So no Hamilton circuit exists.

3. (a) Yes. Try $v_1 v_2 v_6 v_5 v_4 v_3 v_1$, for example.   (b) No.
    (c) No. If v_1 is in V_1 then $\{v_2, v_3, v_4, v_5\} \subseteq V_2$, but v_2 and v_3 are joined by an edge.
    (d) No.

4. (a) No. See part (c) and Theorem 4.   ·   (b) No.
    (c) Yes, with partition $\{\{v_1, v_4, v_7\}, \{v_2, v_3, v_5, v_6\}\}$.   (d) Yes.

5. (a) Since there are $n!$ choices for the order in which the vertices in V_1 and in V_2 are visited and the initial vertex can be in either V_1 or V_2, there are $2(n!)^2$ possible Hamilton circuits.
    (b) $n!\cdot(n-1)!$.
    (c) m and n even, or m odd and $n = 2$, or $m = n = 1$.

6.   (a)   (b)   (c)

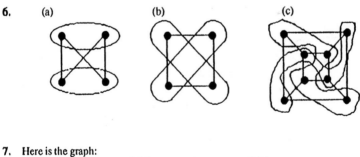

7. Here is the graph:

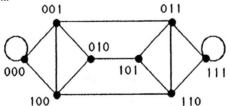

One possible Hamilton circuit has vertex sequence 000, 001, 011, 111, 110, 101, 010, 100, 000 corresponding to the circular arrangement 0 0 0 1 1 1 0 1. Although there are four essentially different Hamilton circuits in $\{0, 1\}^3$, there are only two different circular arrangements, $0\,0\,0\,1\,1\,1\,0\,1$ and $0\,0\,0\,1\,0\,1\,1\,1$, which are just the reverses of each other.

8. The twelve Gray codes of length 3 are the circular lists
   (000, 001, 011, 111, 101, 100, 110, 010),
   (000, 001, 101, 100, 110, 111, 011, 010),
   (000, 100, 101, 001, 011, 111, 110, 010),
   (000, 100, 110, 111, 101, 001, 011, 010),
   (000, 001, 101, 111, 011, 010, 110, 100),
   (000, 001, 011, 010, 110, 111, 101, 100),
   and their reverses.

9. There is no Hamilton path because the graph is not connected. The graph is drawn in the answer to Exercise 9, § 6.2.

10. Yes to both questions. One Hamilton circuit is 000, 011, 100, 111, 001, 010, 101, 110, 000. The graph is drawn in the answer to Exercise 10, § 6.2.

11. (a) K_n^+ has n vertices and just one more edge than K_{n-1} has, so it has exactly
    $\frac{1}{2}(n-1)(n-2) + 1$ edges.
    (b) Consider vertices v and w in K_n^+ which are not connected by an edge. One of them is in K_{n-1}, so it has degree $n-2$ and the other is the new vertex of degree 2. Hence $\deg(v) + \deg(w) = n$, so K_n^+ is Hamiltonian by Theorem 3.

12. Since any circuit through the new vertex must traverse the new edge at least twice, K_n^{++} is not Hamiltonian.

13. (a)   (b) 2.

    (c) Choose two vertices u and v in G. If they are *not* joined by an edge in G, then they are joined by an edge in the complement. If they *are* joined by an edge in G, then they are in the same component of G. Choose w in some other component. Then u w v is a path in the complement. In either case, u and v are joined by a path in the complement.

(d) One example is drawn in the answer to part (e).

(e) No. Consider .

14. Suppose $|V(G)| = n$. Each vertex of the complement of G has degree $n - 1 - k$. Since $n \geq 2k + 2$, $k + 1 \leq n/2$. Thus $n - 1 - k \geq n - n/2 = n/2$. By Theorem 1, G is Hamiltonian.

15. Given G_{n+1}, consider the subgraph H_0 where $V(H_0)$ consists of all binary $(n + 1)$-tuples with 0 in the $(n + 1)$-st digit and $E(H_0)$ is the set of all edges of G_{n+1} connecting vertices in $V(H_0)$. Define H_1 similarly. Show H_0, H_1 are isomorphic to G_n, and so have Hamilton circuits. Use these to construct a Hamilton circuit for G_{n+1}. For $n = 2$, see how this works in Figure 5.

Here are the details. Let $V(H_1)$ consist of $(n+1)$-tuples with 1 in the $(n+1)$-st digit and let $E(H_1)$ consist of the edges in G_{n+1} connecting vertices in $V(H_1)$. The function from G_{n+1} to G_n which simply leaves off the last coordinate determines isomorphisms from H_0 and H_1 onto G_n. Suppose e is an edge from v to w in a Hamilton circuit for G_n and that v_0, w_0 and v_1, w_1 correspond to v, w in H_0 and H_1. I.e., v_0 is the n-tuple v with a 0 attached to the end, etc. Form a Hamilton circuit for G_{n+1} as follows. Starting with w_0 in H_0 follow the copy in H_0 of the Hamilton circuit for G_n until you reach v_0. Then take the edge from v_0 to v_1, which exists since v_0 and v_1 only differ in the $(n+1)$-st coordinate. Then go backwards along the copy of the G_n-circuit in H_1 until you reach w_1. Then take the edge from w_1 to w_0, the starting point. Every vertex in H_0 and H_1 will have been visited exactly once. Since $V(G_{n+1}) = V(H_0) \cup V(H_1)$, the path is a Hamilton circuit of G_{n+1}.

16. The graph G_m has 2^m vertices, each of degree m. For $m \geq 3$, $2^m > m + m$ so G_m does not satisfy the hypothesis $\deg(v) + \deg(w) \geq 2^m$ of Theorem 3 for any pairs of vertices v and w. It follows, as in the proofs of Theorems 1 and 2, that G_m does not satisfy the hypothesis of either of those theorems. So none of the theorems apply to G_m.

Section 6.6

1. (a) and (b)   (c)   (d)

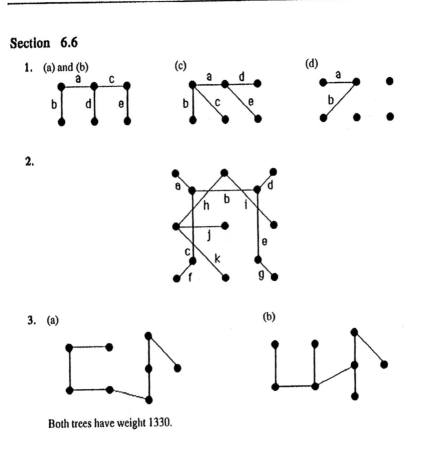

2.

3. (a)   (b)

Both trees have weight 1330.

4.

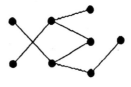

5. (a) and (c)

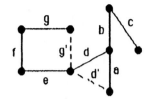

Either d or d' can be chosen, and either g or g', so there are 4 possible answers to (a).

(b) 1330.

6. (a) $e_{10}, e_9, e_7, e_5, e_4, e_3, e_1$.

(b) Same as (a).

7. (a) $e_1, e_2, e_3, e_5, e_6, e_7, e_9$.

(b) $e_7, e_5, e_2, e_1, e_3, e_6, e_9$.

8. (a)

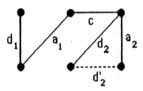

Edges a_1 and a_2 can be chosen in either order. So can d_1 and d_2. Edge d_2' can be chosen instead of d_2. The weight is 10.

(b)

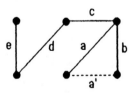

Either a or a' can be chosen. The weight is still 10.

9. (a)

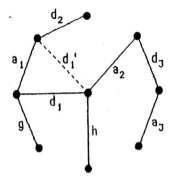

Edges a_1, a_2, a_3 can be chosen in any order. So can d_1, d_2, d_3. Edge d_1' can be chosen instead of d_1. The weight is 16.

(b)

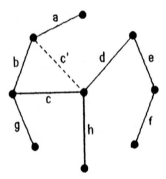

Either c or c' can be chosen. The weight is 16.

10. (a) $\{e_1, e_2, e_3, e_4, e_5\}$, $\{e_1, e_2, e_3, e_4, e_6\}$, $\{e_1, e_2, e_3, e_5, e_6\}$.

(b) e_1, e_2, e_3.

(c) An edge e belongs to every spanning tree of a finite connected graph G if and only if its removal will disconnect the graph. *Proof:* $\Leftarrow$ This is clear since spanning trees must be connected. $\Rightarrow$ We prove the contrapositive, so suppose $G \setminus \{e\}$ is connected. By Theorem 1 in § 6.3, $G \setminus \{e\}$ has a spanning tree, which will be a spanning tree for G that does not use e.

11. 1687 miles.

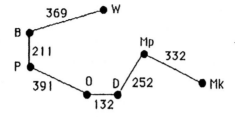

12. Yes. Given an edge, order the edges of the graph so that the given edge is e_1. Then KRUSKAL's algorithm will pick e_1 right off the bat, so e_1 will be part of the spanning tree produced by that algorithm.

13. Here is one possible algorithm:

    Set $E := \emptyset$.

    Choose w in $V(G)$ and set $V := \{w\}$.

    While $V \neq V(G)$

        if there is an edge $\{u, v\}$ in $E(G)$ with $u \in V$ and $v \in V(G) \setminus V$

            choose such an edge of smallest weight

            put $\{u, v\}$ in E and put v in V

        otherwise choose $v \in V(G) \setminus V$ and put v in V. $\square$

14. (a) Consider a minimum spanning tree S of G contained in H. Since $S \subseteq H$, $V(H) = V(S) = V(G)$. Since S is a minimum spanning tree of G, S is a minimum spanning tree of H. Now if T is any minimum spanning tree of H, then $W(T) = W(S)$ and so T is a minimum spanning tree of G.

(b) Here are the details of the outline given. The Kruskal tree K for E is a minimum spanning tree for G by part (a), since E contains some minimum spanning tree for G by assumption. It suffices to show that $K \subseteq E \setminus \{e\}$. Since e is to be deleted, e is in a cycle C of E. Since K is a tree, $C \not\subseteq K$. Consider any edge f of $C \setminus K$ with $f \neq e$. Since f is in a cycle of E and f has not been discarded, f precedes e on the list of edges of G. Since $f \in E \setminus K$, Kruskal's algorithm rejected f; hence f is part of a cycle made from edges in K that precede f, and hence that precede e. That is, the endpoints of f are joined by a path in K of edges preceding e. Replacing each f in $C \setminus K$ by such a path gives a path in $K \setminus \{e\}$ joining the endpoints of e. Since K is a tree, this means that $e \notin K$, and so $K \subseteq E \setminus \{e\}$.

103

15. Assume that G has more than one minimum spanning tree. Consider the edge e of smallest weight that belongs to some but not all minimum spanning trees. Let S, T be minimum spanning trees with $e \in T \setminus S$. By Theorem 2(d) of § 6.3, $S \cup \{e\}$ has a cycle C that must contain e since S is acyclic. Since T is acyclic there must be some other edge, call it f, in C that is not in T. Then f is an edge of S. Now $U = (S \cup \{e\}) \setminus \{f\}$ is connected, by Theorem 3 of § 6.2, and has the same number of edges as S. So U is a spanning tree by Theorem 3 of § 6.3. Since S is a *minimum* spanning tree,

$$\text{weight}(S) \leq \text{weight}(U) = \text{weight}(S) + W(e) - W(f),$$

and so $W(e) \geq W(f)$. Since $e \neq f$, $W(e) > W(f)$. Since $f \in S \setminus T$, this contradicts our choice of e.

CHAPTER 7

Recursion is a fundamental idea in computer science. Our students who have had some experience with recursive programs seem simply to have been told that a recursive algorithm is one that calls itself. Even though it's not hard to put recursion on a respectable mathematical footing, students seem to have just been shown a few examples and told to go and do likewise. In this chapter we describe and relate the various ways in which recursion arises, and we point out what it takes to verify a recursive algorithm. Whether one thinks of such an algorithm as working its way downward or upward, there's generally a tree lurking in the background. We've chosen to discuss algorithms in which the tree is in more or less plain sight, not only to encourage thinking of recursion and trees together but also because the algorithms themselves have important applications.

Section 7.1 presents three views of recursion: recursive definition of sets, a recursive inductive framework, and recursive definition of functions. Recursive algorithms can be thought of as computing values of recursively defined functions; their correctness is guaranteed by the inductive framework. At the root of it all are the recursively defined sets on which the algorithms act. We have tried in this introductory section to give a coherent and convincing discussion of the basic abstract ideas without getting overly technical. What we say is enough for our purposes and should be useful even for those students who go on to study computability questions.

Section 7.2 begins to deal with the algorithms themselves. Testing membership in a recursively defined set is an obvious kind of motivating example. Parsing algorithms for compilers are of this sort. The main theme of this section is the connection between recursive algorithms and the abstract ideas of § 7.1. If you didn't do § 4.6 you can skip over the recursive versions of the Euclidean algorithm in Example 6.

The first part of § 7.3 deals with traversal algorithms whose inputs are trees with labels on their nodes. We'll apply this material to Polish notation in the next section. Go through a couple of examples to illustrate how the algorithms work and to show a format for describing the stages of their operation. Students seem to find all three of these algorithms easy and fun. Their complexity analysis brings in a new and powerful trick, the method of "charges."

The idea of depth-first search is useful even when we don't start with a tree. We illustrate with an algorithm, whose output is interesting in its own right, for producing a topologically sorted labeling of an acyclic digraph. Students find tree traversal pretty easy but LABEL and TREESORT somewhat less intuitive. Work some examples in class for each of these. The idea of keeping track of where we've been with the set L is a new wrinkle that will get used again in Chapter 8. Exercise 14 shows that there are some subtleties in our design of TREESORT. This section is fundamental enough to warrant two class days.

Section 7.4 builds on § 7.3 to discuss an important application. The theorem gives the theoretical justification for the use of Polish and reverse Polish notation. Its proof is slippery. Go through it with an example to see what the notation really means.

Weighted trees have numerous applications. In § 7.5 we discuss two: list-merging and prefix codes. Though we don't explicitly talk about **pruning**, that's in fact what we do in HUFFMAN'S algorithm. Huffman codes are sometimes used for file compression.

Section 7.1

1. (a) Use induction on m. Clearly 2^0 is in S by (B). If 2^m is in S, then $2 \cdot 2^m = 2^{m+1}$ is in S by (R).

    (b) Use the Generalized Principle of Induction, where $p(n) =$ "n has the form 2^m for some $m \in \mathbb{N}$." Statement $p(1)$ is true since $1 = 2^0$. We show that $p(n)$ implies $p(2n)$. If $p(n)$ holds, then $n = 2^m$ for some $m \in \mathbb{N}$. Since $2n = 2^{m+1}$ and $m + 1 \in \mathbb{N}$, statement $p(2n)$ also holds.

2. (a) λ is in Σ^* by (B). $c \in \Sigma$, so $\lambda c = c \in \Sigma^*$ by (R). $a \in \Sigma$, so $ca \in \Sigma^*$ by (R). $t \in \Sigma$, so $cat \in \Sigma^*$ by (R).

    (b), (c), (d) are similar.

3. (a) $S = \{(m,n) : m \leq n\}$.

    (b) By (B), $(0,0) \in S$. So by (R), $(0,1) \in S$. Again by (R), $(1,1) \in S$ so by (R) $(0,2) \in S$. Finally, by (R), $(1,2) \in S$.

    (c) Yes. A pair $(0,n)$ is in S only by (B) or because $(n - 1, n - 1) \in S$, and a pair (m,n) with $0 < m \leq n$ can only come from $(m - 1, n)$.

4. (a) $T = \{(m,n) : m \leq n\}$.

    (b) $(0,0) \in T$ by (B), so $(1,1)$, $(2,2)$, $(3,3)$, $(3,4)$, $(3,5)$ are in T, by repeated use of (R). [Other sequences are possible; for example $(0,0)$, $(0,1)$, $(1,2)$, $(1,3)$, $(2,4)$, $(3,5)$.]

    (c) No, as noted in the answer to part (b). Compare with the answer to Exercise 3(c).

5. (a) (B) λ is in S,   (R) if $w \in S$, then $aw \in S$ and $wb \in S$.

(b) $\lambda \in S$ by (B). Now repeated use of (R) yields $a\lambda \in S$, i.e., $a \in S$, so $ab \in S$, so $abb \in S$, so $abbb \in S$.

(c) $\lambda \in S$ by (B), so a, aa and aab are in S by (R).

(d) Our definition is not uniquely determined. For example, aab can also be built using the sequence λ, b, ab, aab.

6. (a) (B) $a \in T$,   (R) if $w \in T$, then $bw \in T$ and $wb \in T$.

(b) $a \in T$ by (B), so ba, bba and bbab are in T by (R).

(c) Our definition is not uniquely determined. For example, bbab can also be built from the sequence a, ab, bab, bbab.

7. (a) Two possible constructions are

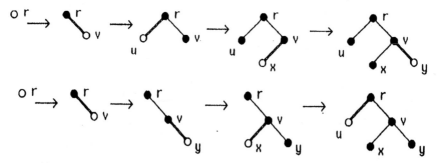

(b) First obtain the subtree with root v by hanging the trivial trees with vertices x and y. Then obtain the tree with root r by hanging the subtree with root v and the trivial tree with vertex u.

8. S_0 consists of the trivial one-vertex tree. Assume that S_{n-1} contains all trees with at most n vertices, and consider a tree T with $n + 1$ vertices. Then T has at least two leaves by Lemma 1, § 6.3. Prune a leaf of T to get a new tree T''. Then T'' belongs to S_{n-1}. Since T can be obtained from T'' by reattaching the leaf, (R) shows that T belongs to the set S_n.

9. Use the Generalized Principle of Induction on $p(n) =$ "$n \equiv 0$ (mod 3) or $n \equiv 1$ (mod 3)." Certainly $p(1)$ is true. It suffices to show that $p(n) \Rightarrow p(3n)$ and $p(2n + 1) \Rightarrow p(n)$. The first implication is trivial because $3n \equiv 0$ (mod 3) for all n. For the second implication, show the contrapositive, i.e., $n \equiv 2$ (mod 3) implies $(2n + 1) \equiv 2$ (mod 3). The contrapositive implication is easy to show directly, since $n = 3k + 2$ implies $2n + 1 = 3(2k + 1) + 2$.

10. The chain in Example 6(a) shows that $9, 13$ and 7 are in A. Since $9 = 2 \cdot 4 + 1$ and $13 = 2 \cdot 6 + 1$ are in A, (R) shows that 4 and 6 are in A. Since $4 \in A$, we have $12 \in A$. Finally, since $7 \in A$ we have $21 \in A$ and hence $10 \in A$.

11. (a) Use (B) and (R) to show that the sequence $1, 2, 4, 8, 16, 5, 10, 3, 6$ lies in S.
    (b) From part (a), 10 is in S. Now use the sequence $10, 20, 40, 13, 26, 52, 17, 34, 11, 22, 7$.

12. (a) Imitate the definition in Example 11(b); in (R) change "at most" to "exactly".
    (b) The class of full m-ary trees is defined by:

    (B) A trivial one-vertex rooted tree is a full m-ary tree of height 0.

    (R) If (T,r) is obtained by hanging exactly m full m-ary rooted trees of height $h - 1$ from r, then (T,r) is a full m-ary tree of height h.

    Note that full m-ary trees are essentially completely determined by giving m and the height h.

13. We follow the hint. Let $p(w) = "\ell'(w)$ is the number of letters in w." By (1) of Example 10(b), $p(w)$ is true if $w \in X = \{\lambda\} \cup \Sigma$. Suppose that $w = uv$ with $p(u)$ and $p(v)$ true. By (2) of Example 10(b), $\ell'(w) = \ell'(u) + \ell'(v) = $ (number of letters in u) + (number of letters in v) = number of letters in w, so $p(w)$ is true. [This argument is valid no matter how w is produced from members of Σ^* by (2).] Thus $p(w)$ is true for every w in Σ^* by the Generalized Principle of Induction.

14. (a) If $x \in \Sigma$, then $x = x\lambda = x\overleftarrow{\lambda}$ [by (B)] $= \overleftarrow{\lambda x}$ [by (R)] $= \overleftarrow{x}$.
    (b) $\overleftarrow{cab} = b\overleftarrow{ca} = ba\overleftarrow{c} = bac$, by two applications of (R) and one application of the result of part (a).
    (c) $\overleftarrow{abbaa} = a\overleftarrow{abba} = aa\overleftarrow{bba} = aab\overleftarrow{ab} = aabb\overleftarrow{a} = aabba$, by repeated use of (R) and part (a).
    (d) $\overleftarrow{w_1 w_2} = \overleftarrow{w_2}\,\overleftarrow{w_1}; \quad \overleftarrow{\overleftarrow{w_1}} = w_1$.

15. (a) $(2,3)$, $(4,6)$, etc.
    (b) (B) is clear since 5 divides $0 + 0$. For (R) you need to check
    "if 5 divides $m + n$, then 5 divides $(m + 2) + (n + 3)$."

    Alternatively, prove that every member of S is of the form $(2k, 3k)$ for $k \in \mathbb{N}$.

    (c) No. For example, $(3,2)$ does not belong to S.

16. (a) T consists of all pairs (m,n) with $n \leq 2m$.

(b) Use the Generalized Principle of Induction with $p(m,n) = $ "$2m \geq n$." Check that $p(0,0)$ is true. Assuming that $2m \geq n$ for some member (m,n) of T, then $2(m + 1) = 2m + 2 \geq n + 2$; also $2(m + 1) \geq n + 1$ and $2(m + 1) \geq n$. So $2x \geq y$ for $(x,y) = (m + 1,n)$, $(m + 1,n + 1)$ and $(m + 1,n + 2)$, i.e., $2x \geq y$ for every pair (x,y) in T specified from (m,n) by (R).

(c) No. Some pairs in T can be defined in terms of more than one pair. For example, $(3,2)$ arises from $(2,2)$, $(2,1)$ and $(2,0)$ under (R).

17. (a) Obviously $A \subseteq \mathbb{N} \times \mathbb{N}$. To show $\mathbb{N} \times \mathbb{N} \subseteq A$, apply the ordinary First Principle of Mathematical Induction to the propositions

$$p(k) = \text{"if } (m,n) \in \mathbb{N} \times \mathbb{N} \text{ and } m + n = k, \text{ then } (m,n) \in A.\text{"}$$

(b) Let $p(m,n)$ be a proposition-valued function defined on $\mathbb{N} \times \mathbb{N}$. To show that $p(m,n)$ is true for all (m,n) in $\mathbb{N} \times \mathbb{N}$ it is enough to show:

(B) $p(0,0)$ is true, and

(I) if $p(m,n)$ is true then $p(m + 1,n)$ and $p(m,n + 1)$ are true.

18. (a) Some examples are a, aba, bab, ababb, babab.

(b) Use the result of Example 9(c) and the Generalized Principle of Induction with $p(w) = $ "length(w) is odd." It is enough to show:

(B) $p(w)$ is true for all $w \in \{a, b\}$;

(I) if $p(w)$ is true for $w \in B$, so are $p(abw)$ and $p(baw)$.

Both of these assertions are clear. (B) holds because length(a) = length(b) = 1. (I) holds because length(abw) = length(baw) = 2 + length(w) for all w.

(c) No. w = aaa has odd length, but $w \notin B$.

(d) Yes. Each word in B of length at least 3 is of just one form abw or baw, and in either case the word w is unique.

19. (a) For w in Σ^*, let $p(w) = $ "length($\overleftarrow{w}$) = length(w)". Apply the Generalized Principle of Induction. Since $\overleftarrow{\lambda} = \lambda$, $p(\lambda)$ is clearly true. You need to show that if $p(w)$ is true, then so is $p(wx)$:

$$\text{length}(\overleftarrow{w}) = \text{length}(w) \text{ implies length}(\overleftarrow{wx}) = \text{length}(wx).$$

In detail, suppose length($\overleftarrow{w}$) = length(w). Then

$$\text{length}(\overleftarrow{wx}) = \text{length}(x\overleftarrow{w})$$
$$= \text{length}(x) + \text{length}(\overleftarrow{w})$$
$$= \text{length}(x) + \text{length}(w) \qquad \text{[by assumption]}$$
$$= \text{length}(wx).$$

(b) Fix w_1, say, and work with $p(w) = "\overleftarrow{w_1 w} = \overleftarrow{w}\overleftarrow{w}_1."$ To show $p(w)$ true for every $w \in \Sigma^*$ it is enough to show:

(B) $p(\lambda)$ is true;

(I) if $p(w)$ is true, then $p(wx)$ is true [for $w \in \Sigma^*$ and $x \in \Sigma$].

Since $\overleftarrow{\lambda} = \lambda$, $\overleftarrow{w_1 \lambda} = \overleftarrow{w}_1 = \overleftarrow{\lambda}\,\overleftarrow{w}_1$ and so (B) holds. If $p(w)$ is true and $x \in \Sigma$ then

$$\overleftarrow{w_1 wx} = x\overleftarrow{w_1 w} \qquad \text{[by definition of reversal]}$$
$$= x\overleftarrow{w}\,\overleftarrow{w}_1 \qquad \text{[by } p(w)\text{]}$$
$$= \overleftarrow{wx}\,\overleftarrow{w}_1 \qquad \text{[by definition of reversal].}$$

Thus $p(wx)$ holds. This establishes (I).

Section 7.2

1. (a) TEST(20;)

      TEST(10;)

         TEST(5;)

           $b := \text{false}; \quad m := -\infty$

         $b := \text{false}; \quad m := -\infty + 1 = -\infty$

      $b := \text{false}; \quad m := -\infty + 1 = -\infty$

(b) TEST(8;)

      TEST(4;)

         TEST(2;)

           TEST(1;)

              $b := \text{true}; \quad m := 0$

           $b := \text{true}; \quad m := 0 + 1 = 1$

         $b := \text{true}; \quad m := 1 + 1 = 2$

      $b := \text{true}; \quad m := 2 + 1 = 3$

2. (a) This is essentially the graph in Figure 1. Prune leaves until a single vertex is reached. Then $b := \text{true}$, i.e., the graph is a tree.

(b) This is essentially the graph H' in Figure 2. Prune leaves until the graph H''' is obtained. Then $b := \text{false}$, so the graph is not a tree.

(c) Prune leaves until a single vertex is left; then $b := \text{true}$ and the graph is a tree.

3. (a) $((x + y) + z)$ or $(x + (y + z))$.

(b) $(x + (y / z))$ or $((x + y) / z)$, or $x + (y / z)$ or $(x + y) / z$ if we omit outside parentheses.

(c) $((xy)z)$ or $(x(yz))$.

(d) $((x + y)^{(x+y)})$ or $(x + y)^{(x+y)}$.

110

4. (a) $(X + Y) + Z$ or $X + (Y + Z)$.   (b) $X * (Y + Z)$.

(c) $(X\hat{\ }2) + ((2 * X) + 1)$ or $((X\hat{\ }2) + (2 * X)) + 1$ or
$((X\hat{\ }2) + 2) * (X + 1)$ or $(X\hat{\ }(2 + 2)) * (X + 1)$ or
$(X\hat{\ }2) + (2 * (X + 1))$ or $X\hat{\ }((2 + (2 * X)) + 1)$ or
$X\hat{\ }(((2 + 2) * X) + 1)$ or $X\hat{\ }((2 + 2) * (X + 1))$ or
$X\hat{\ }(2 + (2 * (X + 1)))$ or $X\hat{\ }(2 + ((2 * X) + 1))$ or
$(X\hat{\ }(2 + (2 * X))) + 1$ or $(X\hat{\ }((2 + 2) * X)) + 1$ or
$(((X\hat{\ }2) + 2) * X) + 1$ or $((X\hat{\ }(2 + 2)) * X) + 1$.

(d) $((X + (Y / Z)) - (Z * X)$ or one of 13 others.

[For simplicity, we have omitted outside parentheses in these answers.]

5. (a) By (B), x, y and 2 are wff's. By the (f^8) part of (R), we conclude that (x^2) and (y^2) are wff's. So by the $(f + g)$ part of (R), $((x^2) + (y^2))$ is a wff.

(b) X, 2 and Y are wff's by (B). $(X\hat{\ }2)$ and $(Y\hat{\ }2)$ are wff's by the $(f\hat{\ }g)$ part of (R). $((X\hat{\ }2) + (Y\hat{\ }2))$ is a wff by the $(f + g)$ part of (R). $(((X\hat{\ }2) + (Y\hat{\ }2))\hat{\ }2)$ is a wff by the $(f\hat{\ }g)$ part of (R).

(c) By (B), X and Y are wff's. By the $(f + g)$ part of (R), $(X + Y)$ is a wff. By the $(f - g)$ part of (R), $(X - Y)$ is a wff. Finally, by the $(f*g)$ part of (R), $((X + Y) * (X - Y))$ is a wff.

6. (a) The algorithm terminates but the output r is always 0.

(b) If the input is n the output is $(n - 1)!$ [recall that $0! = 1$ by definition].

7. Blank entries below signify that the algorithm doesn't terminate with the indicated input n.

n	FOO	GOO	BOO	MOO	TOO	ZOO
8	8	40,320		4	4	8
9	9	362,880			4	8

8. This is similar to Exercise 9. Let p(k) be the statement "TOO produces $k + 1$ whenever $2^k \leq n < 2^{k+1}$." For $k = 0$, this asserts that "TOO produces 1 whenever $n = 1$," which is clear. Assume p(k) is true and consider n where $2^{k+1} \leq n < 2^{k+2}$. Then $2^k \leq n$ DIV 2 $< 2^{k+1}$, so TOO(n DIV 2;) has output $s = k + 1$. The else branch of TOO(n;) sets $r = s + 1 = k + 2$. I.e., TOO produces $k + 2$ whenever $2^{k+1} \leq n < 2^{k+2}$, so that $p(k + 1)$ is true. Hence all statements p(k) are true by induction.

9. Let p(k) be the statement "ZOO produces 2^k whenever $2^k \leq n < 2^{k+1}$." For $k = 0$, this asserts that "ZOO produces 1 whenever $n = 1$," which is clear. Assume p(k) is true

and consider n where $2^{k+1} \leq n < 2^{k+2}$. Then $2^k \leq n \text{ DIV } 2 < 2^{k+1}$, so ZOO(n DIV 2;) has output $s = 2^k$. The else branch of ZOO(n;) sets $r = 2*s = 2*2^k = 2^{k+1}$. I.e., ZOO produces 2^{k+1} whenever $2^{k+1} \leq n < 2^{k+2}$, so that p(k + 1) is true. Hence all statements p(k) are true by induction.

10. EUCLID$^+$(80,35;)
     EUCLID$^+$(35,10;)
          EUCLID$^+$(10,5;)
               EUCLID$^+$(5,0;)
                    $d := 5$; $s := 1$; $t := 0$
               $d := 5$; $s := 0$; $t := 1 - 0 \cdot (10 \text{ DIV } 5) = 1$
          $d := 5$; $s := 1$; $t := 0 - 1 \cdot (35 \text{ DIV } 10) = -3$
     $d := 5$; $s := -3$; $t := 1 - (-3) \cdot (80 \text{ DIV } 35) = 7$.
Check: $80 \cdot (-3) + 35 \cdot 7 = 5$.

11. EUCLID$^+$(108,30;)
     EUCLID$^+$(30,18;)
          EUCLID$^+$(18,12;)
               EUCLID$^+$(12,6:)
                    EUCLID$^+$(6,0;)
                         $d := 6$; $s := 1$; $t := 0$
                    $d := 6$; $s := 0$; $t := 1 - 0 \cdot (12 \text{ DIV } 6) = 1$
               $d := 6$; $s := 1$; $t := 0 - 1 \cdot (18 \text{ DIV } 12) = -1$
          $d := 6$; $s := -1$; $t := 1 - (-1) \cdot (30 \text{ DIV } 18) = 2$
     $d := 6$; $s := 2$; $t = -1 - 2 \cdot (108 \text{ DIV } 30) = -7$.
Sure enough, $108 \cdot 2 + 30 \cdot (-7) = 6$.

12. EUCLID$^+$(56,21;)
     EUCLID$^+$(21,14;)
          EUCLID$^+$(14,7;)
               EUCLID$^+$(7,0;)
                    $d := 7$; $s := 1$; $t := 0$
               $d := 7$; $s := 0$; $t := 1 - 0 \cdot (14 \text{ DIV } 7) = 1$
          $d := 7$; $s := 1$; $t := 0 - 1 \cdot (21 \text{ DIV } 14) = -1$
     $d := 7$; $s := -1$; $t := 1 - (-1) \cdot (56 \text{ DIV } 21) = 3$.
Check: $56 \cdot (-1) + 21 \cdot 3 = 7$.

13. (a) Since gcd(m,n) = m when n = 0, we may assume that n ≠ 0. We need to check that the algorithm terminates and that if the output d' from EUCLID(n,m MOD n;) is correct, then the output d = d' from EUCLID(m,n;) is correct. The latter observation follows from the equality gcd(n,m MOD n) = gcd(m,n), which was verified in § 4.6.

To check that the algorithm terminates, we need to be sure that the second variable n in EUCLID(,n;) is eventually 0. This is clear, since at each step the new n' = m MOD n is less than n and so the values must decrease to 0.

(b) The algorithm terminates because it's the same algorithm as in part (a), but with extra outputs. We need to verify that s·m + t·n = d at each stage of the algorithm. Since this equation holds when n = 0, d = m, s = 1 and t = 0, we may assume n ≠ 0. As in (a), we need to check that if the output of EUCLID⁺(n,m MOD n;) is correct, then so is the output of EUCLID⁺(m,n;). That is, we need to show that

$$d' = s'·n + t'·(m \text{ MOD } n), \quad d = d', \quad s = t', \quad t = s' - t'·(m \text{ DIV } n)$$

all imply that d = s·m + t·n. For this we use the identity m = (m DIV n)·n + (m MOD n):

$$s·m + t·n = t'·m + [s' - t'·(m \text{ DIV } n)]·n$$
$$= t'·[m - (m \text{ DIV } n)·n] + s'·n$$
$$= t'·(m \text{ MOD } n) + s'·n = d' = d.$$

14. (a) 2. (b) 3. (c) 2. (d) 5. (e) 4. (f) 3.

15. (a) p and q are wff's by (B). p ∨ q is a wff by (R). ¬(p ∨ q) is a wff by (R).
(b) p and q are wff's by (B). ¬p and ¬q are wff's by (R). So the expression (¬p ∧ ¬q) is a wff by (R).
(c) p, q and r are wff's by (B). (p ↔ q) and (r → p) are wff's by (R). ((r → p) ∨ q) is a wff by (R), so ((p ↔ q) → ((r → p) ∨ q)) is a wff by (R).

16. Simply add "(P ⊕ Q)," to the recursive clause (R).

17. (a) p,q ∈ 𝓕 by (B). (p ∨ q) ∈ 𝓕 by (R) with P = p, Q = q. Hence (p ∧ (p ∨ q)) ∈ 𝓕 by (R) with P = p, Q = (p ∨ q).
(b) There are too many p's and P's around, so let's use r(P) for the proposition-valued function on 𝓕, to which we wish to apply the general principle of induction. Then you need to prove all r(P) are true where

r(P) = "if p, q are false, then P is false."

To prove r(P) for all P ∈ 𝓕 it is enough to show:
(B) r(p) and r(q) are true;
(I) if r(P) and r(Q) are true, then so are r((P ∧ Q)) and r((P ∨ Q)).

(B) obviously holds. Suppose $r(P)$ and $r(Q)$ are true for some $P,Q \in \mathcal{F}$, and suppose p and q are false. Then P is false, by $r(P)$, and Q is false, by $r(Q)$. Then $P \wedge Q$ and $P \vee Q$ are false by definition of $\wedge$ and $\vee$, so $r((P \wedge Q))$ and $r((P \vee Q))$ are true. Thus (I) holds.

(c) If p and q are false then $(p \rightarrow q)$ is true, so $r((p \rightarrow q))$ is false. Thus $(p \rightarrow q)$ cannot be logically equivalent to a proposition in $\mathcal{F}$, by part (b).

This exercise provides a negative answer to the question in Exercise 17(c) of § 2.4: $p \rightarrow q$ cannot be written in some way using just $p, q, \wedge$ and $\vee$.

Section 7.3

1. Preorder: r x w v y z s u t p q. Postorder: v y w z x t p u q s r.

2. Preorder: r w v x y z u t s p q. Postorder: x y v u t z w p q s r.

3. Preorder: r t x v y z w u p q s. Postorder: v y z x w t p q s u r.

4. 000, 00, 001, 0, 01, root, 10, 1, 1100, 110, 1101, 11, 111.

5. The order of traversing the tree is

r, w, v, w, x, y, x, z, x, w, r, u, t, u, s, p, s, q, s, u, r.

   (a) Preorder: r w v x y z u t s p q. L(w) = w v x y z. L(u) = u t s p q.

   (b) Postorder: v y z x w t p q s u r. L(w) = v y z x w. L(u) = t p q s u.

6.

vertex	1	2	3	4	5	6	7	8
SUCC( )	∅	{1}	∅	{1,3}	{1,2,4}	{5}	{2}	{5,7}
ACC( )	{1}	{1,2}	{3}	{1,3,4}	{1,2,3,4,5}	{1,2,3,4,5,6}	{1,2,7}	{1,2,3,4,5,7,8}

7. The order of traversing the tree is

u, x, w, v, w, y, w, x, r, z, r, t, r, x, u, s, p, s, q, s, u.

   (a) Postorder: v y w z t r x p q s u.

   (b) Preorder: u x w v y r z t s p q.

   (c) Inorder: v w y x z r t u p s q.

8. The tree in Figure 3 and the ones in Figure 4 all have inorder listing v w y x z r t u p s q. The trees in Figure 3 and Figure 5(a) both have preorder listing r w v x y z u t s p q. The trees in Figure 3 and Figure 5(b) both have postorder listing v y z x w t p q s u r.

9. (a)

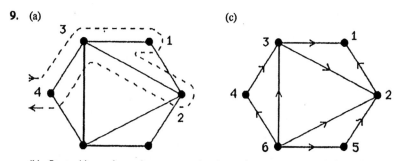

(c)

(b) Start with a, choose its successor b, choose its successor c, label c with 1.
Return to b, choose its successor d and label d with 2.
Return to b and label b with 3.
Return to a and label a with 4.

10. (a)

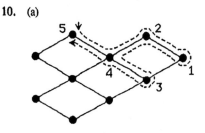

(b)

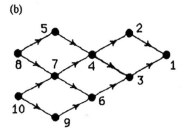

11. (a)

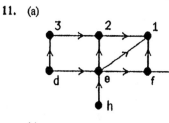

(b

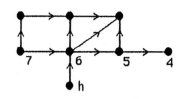

(c)

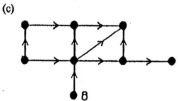

12.

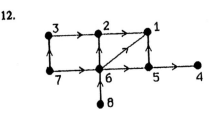

13. (a)

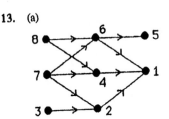

(b)

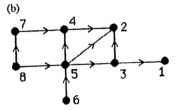

14. (a) The labeling is

(b) Neither the labeling nor its reverse is sorted. Note that $2 < 3 < 4$ but that d is a successor of both b and c.

15. (a) The base cases are the ones in which the tree consists just of the root r.

(b) $n = k + 1$.

(c) The number of descendants of v is a good measure. Each child w of v has fewer descendants than v has. The measure of a base case is 0.

16. (b) There are exactly $n(n - 1)/2$ pairs of distinct vertices, so for each pair (u,v) there is either an edge from u to v or one from v to u [but not both, by acyclicity]. If every vertex has an edge leading from it, then by following edges we construct a closed path, and hence a cycle. Thus some vertex s has no edge pointing away from it. But then every other vertex has an edge to s, so s must be labeled 1. Remove s and the $n - 1$ edges leading to it and apply induction.

17. Since the digraph is acyclic there can be at most one edge joining each of the $n(n - 1)/2$ pairs of distinct vertices.

Section 7.4

1. Reverse Polish: $x\,4\,2\,\hat{}\,-\,y*2\,3\,/\,+$. Polish: $+*-x\,\hat{}\,4\,2\,y\,/\,2\,3$.

2. Reverse Polish: $2\,a\,\hat{}\,2\,b\,\hat{}\,+$. Infix: $(2\,\hat{}\,a)+(2\,\hat{}\,b)$, i.e., 2^a+2^b.

3. (a) Polish: $-*+a\,b-a\,b-\hat{}\,a\,2\,\hat{}\,b\,2$;
   Infix: $(a+b)*(a-b)-((a\,\hat{}\,2)-(b\,\hat{}\,2))$.
   (b) $(a+b)(a-b)-(a^2-b^2)=0$. The tree is trivial.

4. (a) 22.   (b) 49.   (c) 37.

5. (a) 20.   (b) 10.

6. (a) 73.   (b) 1.   (c) 223.   (d) $14/49=2/7$.   (e) 20.

7. (a) $3\,x*4-2\,\hat{}$.   (b) $a\,2\,b*+a\,2\,b*-/$.
   (c) The answer depends on how the terms are associated. For the choice $(x-x^2)+(x^3-x^4)$, the answer is $x\,x\,2\,\hat{}\,-x\,3\,\hat{}\,x\,4\,\hat{}\,-+$.

8. (a) $\hat{}\,-*3\,x\,4\,2$.   (b) $/+a*2\,b-a*2\,b$.
   (c) As in 7(c), the answer depends on how the terms are associated. For the grouping $((x-x^2)+x^3)-x^4$, the answer is $-+-x\,\hat{}\,x\,2\,\hat{}\,x\,3\,\hat{}\,x\,4$.

9. (a) $a\,b\,c**$ and $a\,b*c*$.
   (b) $a\,b\,c+*$ and $a\,b*a\,c*+$.
   (c) The associative law is $a\,b\,c** = a\,b*c*$. The distributive law is
   $$a\,b\,c+* = a\,b*a\,c*+.$$

10. $[(x+y)^2-(x-y)^2]/(xy)=4$.

11. (a) $p\rightarrow(q\vee(\neg p))$.
    (b) Reverse Polish: $p\,q\,p\,\neg\,\vee\rightarrow$; Polish: $\rightarrow p\vee q\,\neg\,p$.

12. (a) Infix: $\neg((\neg p)\wedge(\neg q))\leftrightarrow(p\vee q)$.
    (b) Infix: $(p\wedge q)\leftrightarrow\neg(p\rightarrow\neg q)$.

117

13. (a) Infix: $(p \wedge (p \rightarrow q)) \rightarrow q$.

(b) Infix: $\{[(p \rightarrow q) \wedge (r \rightarrow s)] \wedge (p \vee r)\} \rightarrow (q \vee s)$.

14. (a) $p\,q \rightarrow q\,r \rightarrow \wedge p\,r \rightarrow \rightarrow$.

(b), $p\,q \vee p \neg \wedge q \rightarrow$.

15. (a) Both give $a\,/\,b + c$.

(b) Both give $a + b \hat{\ } 3 + c$.

16. (a) By (B), each of $3, x$ and 2 is a wff. So by (R), $x\,2\hat{\ }$ is a wff. So by (R), $3\,x\,2\hat{\ }*$ is.

(b) By (B), each of x, y and 1 is a wff. So by (R), $x\,y\,+$, $1\,x\,/$ and $1\,y\,/$ are wff's. So $1\,x\,/\,1\,y\,/\,+$ is a wff. Finally $x\,y\,+\,1\,x\,/\,1\,y\,/\,+\,*$ is a wff.

(c) By (B), each of $4, x, 2, y$ and z is a wff. By (R), $x\,2\hat{\ }$ and $y\,z\,+$ are wff's. So $y\,z\,+\,2\hat{\ }$ is a wff. So $x\,2\hat{\ }y\,z\,+\,2\hat{\ }/$ is a wff. Finally, $4\,x\,2\hat{\ }y\,z\,+\,2\hat{\ }/\,-$ is a wff.

17. (a) (B) Numerical constants and variables are wff's.

(R) If f and g are wff's, so are $+f\,g$, $-f\,g$, $*f\,g$, $/f\,g$ and $\hat{\ }f\,g$.

(b) By (B), each of $x, 4$ and 2 is a wff. By (R), $/\,4\,x$ is a wff. So by (R), $+\,x\,/\,4\,x$ is a wff and then $\hat{\ }+\,x\,/\,4\,x\,2$ is a wff.

18. (a) By (B), the variables $x_1, x_2, \ldots$ and the constant 2 are wff's. By (R), $S_1 = x_1\,2\hat{\ }$ is a wff. If S_n is a wff, then by (R) so are $x_{n+1}\,2\hat{\ }$ and $S_{n+1} = S_n\,x_{n+1}\,2\hat{\ }+$. So all S_n's are wff's by induction.

(b) $S_n = x_1^2 + \cdots + x_n^2$ or $\sum_{k=1}^{n} x_k^2$.

19. (a) (B) Variables, such as p, q, r, are wff's.

(R) If P and Q are wff's, so are $P\,Q\,\vee$, $P\,Q\,\wedge$, $P\,Q\,\rightarrow$, $P\,Q\,\leftrightarrow$ and $P\,\neg$.

(b) Argue, in turn, that $q\,\neg$, $p\,q\,\neg\,\wedge$ and $p\,q\,\neg\,\wedge\,\neg$ are wff's. Likewise, $p\,q\,\neg\,\rightarrow$ is a wff. Thus $p\,q\,\neg\,\wedge\,\neg\,p\,q\,\neg\,\rightarrow\,\vee$ is a wff.

(c) (B) Variables, such as p, q, r, are wff's.

(R) If P and Q are wff's, so are $\vee\,P\,Q$, $\wedge\,P\,Q$, $\rightarrow\,P\,Q$, $\leftrightarrow\,P\,Q$ and $\neg\,P$.

(d) By (B), p and q are wff's. By (R), $\neg\,q$ is a wff. So by (R), $\wedge\,p\,\neg\,q$ and $\rightarrow\,p\,\neg\,q$ are wff's. Also $\neg\,\wedge\,p\,\neg\,q$ is a wff. Finally, $\vee\,\neg\,\wedge\,p\,\neg\,q\,\rightarrow\,p\,\neg\,q$ is a wff.

20. (a)

vertex	s	t	v	y	r	z	w	u	x	q
available children	Ø	{s}	{s,t}	{s,t,v}	{s,y}	{r}	{r,z}	{r,z,w}	{r,z,w,u}	{r,z,x}
assigned children				{t,v}	{s,y}				{w,u}	{r,z,x}

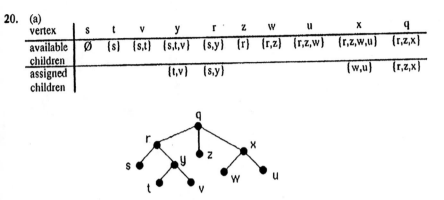

(b) No. s would have to be the root and have no children. Also, the last vertex q would have to be a leaf; it couldn't have 3 children.

Section 7.5

1. (a) 35, 56, 70, 82. (b) 59, 95, 118, 135, 145, 150.

2. Both have weight 150.

3. We recommend the procedure in Examples 4 and 5.

(a) 1,3,4,6,9,13 → 4,4,6,9,13 → 6,8,9,13 → 9,13,14 → 14,22. Weight = 84.

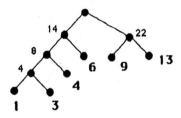

(b) Weight = 136. (c) Weight = 244.

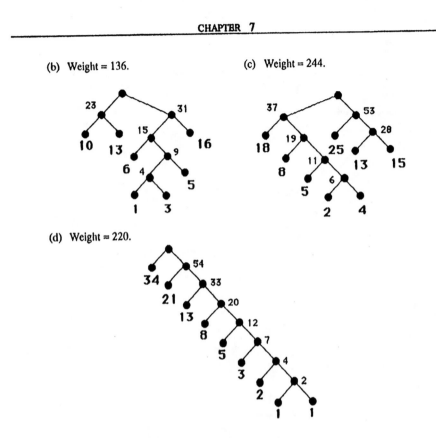

(d) Weight = 220.

4. You will get essentially the tree in Figure 9(a) or the shorter tree

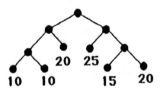

5. All but (b) are prefix codes. In (b), 01 consists of the first two digits of 0111.

6. (a)

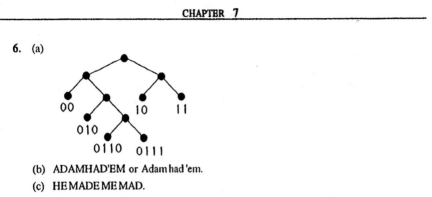

(b) ADAMHAD'EM or Adam had 'em.

(c) HE MADE ME MAD.

(d) DAD HAD MADAM. HE MADE MA MAD. MA MADE DAD DEAD.

7. (a)

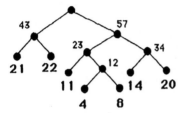

The corresponding optimal code is

letter	a	b	c	d	e	f	g
frequency	11	20	4	22	14	8	21
code	100	111	1010	01	110	1011	00

(b) 269.

8. (a)

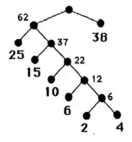

Optimal code

letter	a	b	c	d	e	f	g
frequency	25	2	15	10	38	4	6
code	00	011110	010	0110	1	011111	01110

(b) Average length per 100 letters: 239.

9. (a) The vertex labeled 0 has only one child, 00. See figure below.
   (b) Consider any string beginning with 01.

(a)   (c)

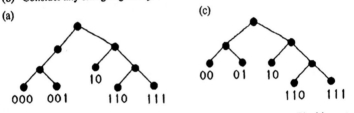

In both (a) and (c), no labeled vertex lies below another such vertex. The binary tree in (c) is regular.

10. (a)

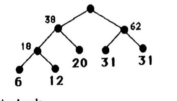

Optimal code

letter	a	d	e	h	m
frequency	31	31	12	6	20
code	10	11	001	000	01

(b) Average length per 100 letters: 218.

11. (a) $(61 + 73 - 1) + (31 + 61 + 73 - 1) + (23 + 31 + 61 + 73 - 1)$
    $= 133 + 164 + 187 = 484.$
   (b) $(23 + 73 - 1) + (23 + 61 + 73 - 1) + 187 = 95 + 156 + 187 = 438.$
   (c) $(23 + 31 - 1) + (61 + 73 - 1) + 187 = 53 + 133 + 187 = 373.$
   (d) $(23 + 73 - 1) + (31 + 61 - 1) + 187 = 95 + 91 + 187 = 373.$

(e)

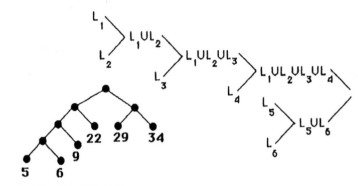

This merging involves at most $(23 + 31 - 1) + (23 + 31 + 61 - 1) + 187 = 53 + 114 + 187 = 354$ comparisons.

12. (a)

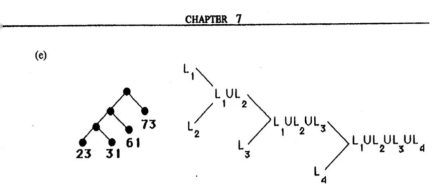

(b) $10 + 19 + 41 + 62 + 104 = 236.$

13. (a)

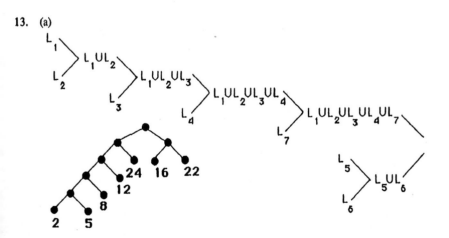

(b) 221.

14. This follows immediately from the lemma after Example 8.

15. For example, in Exercise 1(a), the first tree has weight 35. The leaf of weight $21 = 12 + 9$ was replaced by the subtree with weights 12 and 9 and the weight of the whole tree increased to 56, i.e., to $35 + 21$. Then the leaf of weight $14 = 7 + 7$ was replaced by the subtree with weights 7 and 7 and the weight of the whole tree increased to 70, i.e., to $56 + 14$. Etc.

CHAPTER 8

Some of the material in this chapter is just digraph versions of results about graphs in Chapter 6, but most of it has a flavor that comes from the fact that edges are directed. Acyclic graphs are just forests of trees, but acyclic digraphs can have all kinds of complicated cross-connections.

Section 8.1 introduces sources, sinks and the reachability relation. The algorithm for numbering vertices based on sinks is easy to understand but not as fast as TREESORT in Chapter 7.

In § 8.2 we put weights on the edges, much as we did for graphs in § 6.6. The tables of edge weights W and minweights W^* lead naturally to the use of matrices in algorithms to find W^* and do a variety of other chores. Here we explicitly consider paths of length 0. Later we'll see a way to ignore them selectively. The discussion of scheduling networks can be deemphasized, if desired. The basic ideas are just perversions of the min-weight ideas, but the application itself is of considerable practical importance. Critical path software is now available for microcomputers.

Section 8.3 presents several algorithms for computing min-weights and max-weights. DIJKSTRA's algorithm spreads out from an initial vertex v_0, and when it labels a new vertex v it also computes the min-weight $W^*(v_0,v)$. WARSHALL's algorithm, on the other hand, only guarantees the right answer at the last iteration. Go through examples of both algorithms in class. The format we have used to illustrate the progress of DIJKSTRA's algorithm in Figure 2 should be supplemented in class by marking the graph itself as the labeling proceeds. WARSHALL's algorithm is harder to draw pictures of, though easier to understand; it looks first for paths with just 1 as intermediate vertex, then for those with 1 or 2 as intermediate vertices, then for those with $1, 2$ or 3, etc.

One suggestion for WARSHALL'S algorithm is to distribute sheets on which the algorithm is completely carried out, and discuss a few sample calculations. Or you can just display an intermediate matrix and discuss how to compute entries for the next matrix. Examples are essential. The easiest way to understand how MAX-WEIGHT works is to walk through a simple example. Exercises 1-9 are chosen to provide easy illustrations, with no tricks.

Section 8.4 is a short section that builds on § 8.3, using pointers and linked lists to construct min-paths and max-paths. It also notes how to use the new algorithms to answer other graph-theoretic questions. Some students have trouble grasping the idea of a pointer; a few worked examples in class are important.

Section 8.1

1. Sinks are t and z. Only source is u.

2. (a) SUCC((t) = Ø, SUCC((u) = {t, w, x}, SUCC((v) = {t, y}, SUCC((w) = {y}, etc.
   (b) z.   (c) t and z.

3. (a) R(s) = {s, t, u, w, x, y, z} = R(t), R(u) = {w, x, y, z}, R(w) = {z}, R(x) = Ø, R(y) = {x, z}, R(z) = Ø.
   (b) x and z are the sinks.
   (c) s t s is a cycle, so G is not acyclic.

4. No. It could go around a cycle forever and never stop.

5. (a)

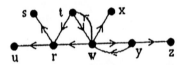

   (b) No, though the immediate successor sets do determine a digraph if multiple edges are not allowed.
   (c) s, u, x, z.
   (d) w x, w r s, w r u, w y z are some.

6. Use NUMBERING VERTICES. One example labels v = 1, u = 2, y = 3, x = 4, w = 5, t = 6, z = 7.

7. Use NUMBERING VERTICES. One example labels t = 1, z = 2, y = 3, w = 4, v = 5, x = 6, u = 7.

8. (a) Here is one. The arrows can be given any directions.

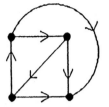

(b) If x and y are two vertices there is either an edge from x to y, so that x is not a sink, or an edge from y to x, so that y is not a sink.

(c) Yes. See the answer to part (a).

(d) Not very much.

9. (a) One example is r s w t v w v u x y z v r u y v s r.

(b) One example is w w x y x z z y w.

10. One might use the sequence $e_2, e_7, e_{13}, e_{12}, e_{11}, e_{14}, e_{16}, e_{15}, e_{10}, e_4, e_5, e_8, e_9, e_6, e_3, e_1$ to get 1 1 0 1 1 1 1 0 0 1 0 1 0 0 0 0 in a circle. Another might use $e_2, e_5, e_6, e_4, e_7, e_{13}, e_{12}, e_9, e_8, e_{11}, e_{14}, e_{16}, e_{15}, e_{10}, e_3, e_1$ to get 1 0 0 1 1 0 1 0 1 1 1 1 0 0 0 0 in a circle.

11. (a) One such digraph is drawn in Figure 3(b) where $w = 0\,0$, $x = 0\,1$, $z = 1\,1$, and $y = 1\,0$.

(b) One possible sequence is 1 1 1 0 1 0 0 0 placed in a circle.

12. (a) The set of 2^{n-1} vertices consists of all strings of 0's and 1's of length $n - 1$. A directed edge connects two such strings if the last $n - 2$ digits of the initial vertex agree with the first $n - 2$ digits of the terminal vertex. Label each edge with the last digit of the terminal vertex. An easy induction shows that for $k = 1, \ldots, n - 1$, the labels of the edges of a path of length k give the last k digits of the terminal vertex.

Now consider any sequence $d_1 d_2 \cdots d_n$ of 0's and 1's. Then $d_1 d_2 \cdots d_{n-1}$ is the initial vertex for an edge labeled d_n. Since the $n - 1$ edges preceding it are labeled $d_1, \ldots, d_{n-1}$, the n consecutive edges are labeled $d_1, \ldots, d_n$, as desired.

(b) Put 0 0 1 1 in a circle.

13. Show that $\hat{G}$ is also acyclic. Apply Theorem 2 to $\hat{G}$. A sink for $\hat{G}$ is a source for G.

127

14. (a) Here is one possible algorithm.

    {input: a finite acyclic digraph G with n vertices}

    {output: a sorted labeling of the vertices of G}

    Let $V := V(G)$ and $E := E(G)$.

    While $V \neq \emptyset$

        let H be the digraph with vertex set V and edge set E

        apply SOURCE to H {to get a source for H}

        label SOURCE(H) with |V|

        remove SOURCE(H) from V and all edges attached to it from E. ☐

(b)

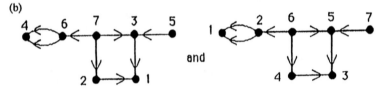

are two possible answers.

15 (a) See the second proof of Theorem 2.

(b) If a finite acyclic digraph has just one source, then there is a path to each vertex from the source.

16. (a) The reflexive and symmetric properties are built into the definition. Transitivity is a general property of reachability: string together a path from u to v and a path from v to w to get a path from u to w.

(b) {s, t}, {u}, {w}, {x}, {y}, {z}.

(c) It is the equality relation: $x \sim y$ if and only if $x = y$. To see this, suppose $x \sim y$ and $x \neq y$. Then there is a path from x to y and a path from y to x; together they give a closed path and Corollary 1 to Theorem 1 shows that G has a cycle.

17. (a) In the proof of Theorem 1 [given in § 6.1], choose a shortest path consisting of edges of the given path.

(b) If $u \neq v$, then Theorem 1 guarantees an acyclic path from u to v and Corollary 2 says such a path has no repeated vertices, so it surely has no repeated edges. If $u = v$, then Corollary 1 says there is a cycle from u to u. Again, all vertices are different, so all edges are too.

18. (a) Consider $w \in R(u)$. Then there is a path from u to w. Since there is a path from v to u there is one from v to w; i.e., $w \in R(v)$. Thus $R(u) \subseteq R(v)$.

(b) Consider a finite acyclic digraph G, and choose v in V(G) with |R(v)| as small as possible. We claim that R(v) = Ø, so that v is a sink. Suppose w ∈ R(v). By part (a), R(w) ⊆ R(v). Since |R(v)| is minimal, R(w) = R(v). Thus w ∈ R(w). By Corollary 1 of Theorem 1, G contains a cycle, contrary to hypothesis.

(c) Yes. Given a list of the sets R(v), one could examine them one at a time to see which has smallest size. In fact, the smallest ones will be empty.

Section 8.2

1.

W	A	B	C	D
A	1.4	1.0	∞	∞
B	0.4	7	∞	0.2
C	0.7	0.3	7	∞
D	0.8	∞	0.2	7

W*	A	B	C	D
A	∞	1.0	1.4	1.2
B	0.4	∞	0.4	0.2
C	0.7	0.3	∞	0.5
D	0.8	0.5	0.2	∞

2.

W*	s	u	v	w	x	y	f
s	0	2	7	4	5	8	9
u	∞	0	∞	2	3	6	7
v	∞	∞	0	∞	∞	2	5
w	∞	∞	∞	0	∞	4	8
x	∞	∞	∞	∞	0	6	4
y	∞	∞	∞	∞	∞	0	4
f	∞	∞	∞	∞	∞	∞	0

3.

W	m	q	r	s	w	x	y	z
m	∞	6	∞	2	∞	4	∞	∞
q	∞	∞	4	∞	4	∞	∞	∞
r	∞	∞	∞	∞	∞	∞	∞	3
s	∞	3	∞	∞	5	1	∞	∞
w	∞	∞	2	∞	∞	∞	2	5
x	∞	∞	∞	∞	3	∞	6	∞
y	∞	∞	∞	∞	∞	∞	∞	1
z	∞	∞	∞	∞	∞	∞	∞	∞

W*	m	q	r	s	w	x	y	z
m	∞	5	8	2	6	3	8	9
q	∞	∞	4	∞	4	∞	6	7
r	∞	∞	∞	∞	∞	∞	∞	3
s	∞	3	6	∞	4	1	6	7
w	∞	∞	2	∞	∞	∞	2	3
x	∞	∞	5	∞	3	∞	5	6
y	∞	∞	∞	∞	∞	∞	∞	1
z	∞	∞	∞	∞	∞	∞	∞	∞

4. s w v y x f is the other min-path.

5. (a) If the digraph is acyclic the weights must be as shown

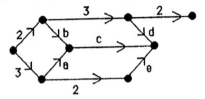

with min{a + 1, b} = 4 and min{c + 1, d, e} = 3.
(b) They might. For example, consider b = 8, a = 3 in part (a), and then consider b = 4, a = 7.

6. String together a min-path from u to v and one from v to w. The weight of the combined path is $W^*(u,v) + W^*(v,w) = W^*(u,w)$, so the path has minimum weight for paths from u to w.

7. (a) The critical paths are s v x f and s v x z f.
(b) If the edges all have positive weights, then going around a cycle again and again would give arbitrarily large path weights.

8. $F(s,z) = L(z) - A(s) - W(s,z) = 104 - 0 - 20 = 84$. Also $F(z,f) = 84$.

9. (a)

	s	u	v	w	x	y	f
A	0	2	7	5	5	11	15
L	0	2	9	7	5	11	15

(b) S(v) = S(w) = 2. S(t) = 0 for all other vertices t.
(c) s u x y f is the only critical path.
(d) Edges on the critical path have float time 0. Also $F(x,f) = 6$, $F(u,w) = F(v,f) = 3$ and $F(s,w) = F(w,y) = F(s,v) = F(v,y) = 2$.

10. (a)

	s	u	v	w	x	y	f
A	0	1	1	2	2	3	4
L	0	1	2	2	2	3	4

(b) S(v) = 1. S(t) = 0 for all other vertices t.

(c) The critical paths are s u w y f and s u x y f.

(d) Edges on the critical paths have float time 0. Also $F(s,w) = F(s,v) = F(v,y) = F(x,f) = 1$ and $F(v,f) = 2$.

11. (a)

	m	s	q	x	w	r	y	z
A	0	2	6	4	10	12	12	15
L	0	3	6	7	10	12	14	15

(b) $S(s) = 1$, $S(x) = 3$, $S(y) = 2$ and $S(t) = 0$ for all other vertices t.

(c) There are two critical paths: m q w z and m q w r z.

(d) Edges on the critical paths have float time 0. Also $F(s,x) = F(x,y) = 4$, $F(s,w) = F(x,w) = F(m,x) = 3$, $F(q,r) = F(w,y) = F(y,z) = 2$ and $F(s,q) = F(m,s) = 1$.

12. (a) $F(s,f) = 0$ and $F(s,u) = F(u,v) = F(v,f) = 6$.

(b) No. If each noncritical task in part (a) were delayed by its float time 6, the total time would be $1 + 2 + 1 + 3 \cdot 6 = 22$.

13. (a) The two critical paths are s u w x y f and s t w x y f. The critical edges are (s,u), (u,w), (s,t), (t,w), (w,x), (x,y) and (y,f).

(b) 2.   (c) The edges are (u,v) and (x,z).

14.

	s	r	t	u	v	w	x	y	z	f
A	0	1	2	3	5	4	9	11	11	14
L	0	2	2	3	6	4	9	11	12	14
S	0	1	0	0	1	0	0	0	1	0

15. (a) Shrink the 0-edges to make their two endpoints the same.

(b)

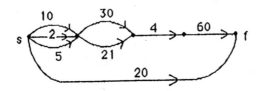

16. (a)

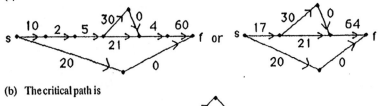

(b) The critical path is

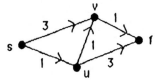

(c) Only frying and rice cooking are not critical.

17. (a)

W	u	v	w	x	y
u	∞	1	∞	∞	∞
v	∞	∞	3	-2	∞
w	∞	∞	∞	∞	∞
x	4	∞	∞	5	∞
y	∞	∞	∞	5	∞

W*	u	v	w	x	y
u	3	1	4	-1	∞
v	2	3	3	-2	∞
w	∞	∞	∞	∞	∞
x	4	5	8	3	∞
y	9	10	13	5	∞

(b) The diagonal entries of W^* are not all ∞'s, so the digraph contains a cycle. [It is, of course, u v x u.]

(c) There would be no min-weights at all for paths involving u, v or x, because going around the cycle u v x u repeatedly would keep reducing the weight by 1.

(d) The sources are vertices whose columns are all ∞'s, and the sinks have rows all ∞'s. The only source is y; the only sink is w.

18. (a) The following are equivalent to $S(v) \leq F(u,v)$:
$L(v) - A(v) \leq L(v) - A(u) - W(u,v)$; $A(u) + W(u,v) \leq A(v)$;
$M(s,u) + W(u,v) \leq M(s,v)$. The last inequality is clear.

(b) If (u,v) is a critical edge, then $F(u,v) = 0$ by (a) of the theorem. Then (b) of the same theorem shows that $S(u) = S(v) = 0$.

(c) (u,v) might not be critical. Here is an example.

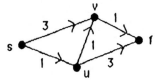

19. (a) $FF(u,v) = A(v) - A(u) - W(u,v)$.

(b) $FF(s,u) = FF(u,x) = FF(x,y) = FF(s,w) = FF(y,f) = FF(s,v) = 0$, $FF(u,w) = 1$,
$FF(w,y) = FF(v,y) = 2$, $FF(v,f) = 3$, $FF(x,f) = 6$.

(c) The slack time at v.

20. (a) Consider a critical edge (u,v). There is a max-path from s to f that contains this edge and has weight $M(s,f)$. The same path with the added weight on (u,v) has total weight greater than $M(s,f)$.

(b) If there is a critical path that does not involve the step, no reduction occurs. For example, see the network in the answer to Exercise 18(c); reducing the weight of edge (s,v) will not reduce the total time needed.

21. (a) $A(u) = M(s,u)$ = weight of a max-path from s to u. If there is an edge (w,u), a max-path from s to w followed by that edge has total weight at most $M(s,u)$. That is, $A(w) + W(w,u) \leq A(u)$. If (w,u) is an edge in a max-path from s to u, then $A(w) + W(w,u) = A(u)$.

(b) Consider a vertex u. If (u,v) is an edge, then (u,v) followed by a max-path from v to f has total weight at most $M(u,f)$. That is, $W(u,v) + M(v,f) \leq M(u,f)$. Hence
$$L(v) - W(u,v) = M(s,f) - M(v,f) - W(u,v) \geq M(s,f) - M(u,f) = L(u)$$
for every edge (u,v), and so $\min\{L(v) - W(u,v) : (u,v) \in E(G)\} \geq L(u)$. Choosing an edge (u,v) in a max-path from u to f gives $L(v) - W(u,v) = L(u)$, so $L(u)$ is the minimum value.

Section 8.3

1. (a)

$$W^* = \begin{bmatrix} \infty & 1 & 2 & 3 & 4 & 5 & 7 \\ \infty & \infty & \infty & 4 & 3 & 4 & 6 \\ \infty & \infty & \infty & 1 & 4 & 5 & 7 \\ \infty & \infty & \infty & \infty & 3 & 4 & 6 \\ \infty & \infty & \infty & \infty & \infty & 1 & 3 \\ \infty & \infty & \infty & \infty & \infty & \infty & 3 \\ \infty & \infty & \infty & \infty & \infty & \infty & \infty \end{bmatrix}.$$

(b)

$$W^* = \begin{bmatrix} \infty & 6 & 8 & \infty & \infty & 1 & 12 \\ \infty & 14 & 2 & \infty & \infty & \infty & 6 \\ \infty & 12 & 14 & \infty & \infty & \infty & 4 \\ 3 & 9 & 11 & \infty & 9 & 4 & 15 \\ 2 & 7 & 9 & \infty & \infty & 3 & 13 \\ \infty & 5 & 7 & \infty & \infty & \infty & 11 \\ \infty & 8 & 10 & \infty & \infty & \infty & 14 \end{bmatrix}.$$

2.

$$M = \begin{bmatrix} -\infty & 1 & 2 & 5 & 8 & 13 & 16 \\ -\infty & -\infty & -\infty & 4 & 7 & 12 & 15 \\ -\infty & -\infty & -\infty & 1 & 5 & 9 & 12 \\ -\infty & -\infty & -\infty & -\infty & 3 & 8 & 11 \\ -\infty & -\infty & -\infty & -\infty & -\infty & 1 & 4 \\ -\infty & -\infty & -\infty & -\infty & -\infty & -\infty & 3 \\ -\infty & -\infty & -\infty & -\infty & -\infty & -\infty & -\infty \end{bmatrix}.$$

3. (a)

L	D(2)	D(3)	D(4)	D(5)	D(6)	D(7)
∅	1	2	∞	∞	∞	∞
{2}	1	2	5	4	∞	∞
{2, 3}	1	2	3	4	9	∞
{2, 3, 4}	1	2	3	4	9	∞
{2, 3, 4, 5}	1	2	3	4	5	7
{2, 3, 4, 5, 6}	1	2	3	4	5	7

no change now.

(b)

L	D(2)	D(3)	D(4)	D(5)	D(6)	D(7)
∅	7	∞	∞	∞	1	∞
{6}	6	10	∞	∞	1	∞
{6, 2}	6	8	∞	∞	1	∞
{6, 2, 3}	6	8	∞	∞	1	12
{6, 2, 3, 7}	6	8	∞	∞	1	12

no change now.

4.

L	D(2)	D(3)	D(4)	D(5)	D(6)
∅	7	∞	2	∞	∞
{4}	6	∞	2	∞	∞
{4, 2}	6	10	2	7	∞
{4, 2, 5}	6	9	2	7	∞
{4, 2, 5, 3}	6	9	2	7	12

The final values of $D(2), \ldots , D(6)$ agree with the values $W^*[1,2], \ldots , W^*[1,6]$ in Example 2.

5. (a)

$$W_2 = \begin{bmatrix} \infty & \infty & \infty & \infty & 1 & \infty & \infty \\ \infty & \infty & \infty & \infty & \infty & \infty & 1 \\ \infty & \infty & \infty & 1 & \infty & 1 & \infty \\ \infty & \infty & 1 & \infty & 1 & \infty & \infty \\ 1 & \infty & \infty & 1 & 2 & \infty & \infty \\ \infty & \infty & 1 & \infty & \infty & \infty & 1 \\ \infty & 1 & \infty & \infty & \infty & 1 & 2 \end{bmatrix}, \qquad W_4 = \begin{bmatrix} \infty & \infty & \infty & \infty & 1 & \infty & \infty \\ \infty & \infty & \infty & \infty & \infty & \infty & 1 \\ \infty & \infty & 2 & 1 & 2 & 1 & \infty \\ \infty & \infty & 1 & 2 & 1 & 2 & \infty \\ 1 & \infty & 2 & 1 & 2 & 3 & \infty \\ \infty & \infty & 1 & 2 & 3 & 2 & 1 \\ \infty & 1 & \infty & \infty & \infty & 1 & 2 \end{bmatrix},$$

$$W_7 = \begin{bmatrix} 2 & 6 & 3 & 2 & 1 & 4 & 5 \\ 6 & 2 & 3 & 4 & 5 & 2 & 1 \\ 3 & 3 & 2 & 1 & 2 & 1 & 2 \\ 2 & 4 & 1 & 2 & 1 & 2 & 3 \\ 1 & 5 & 2 & 1 & 2 & 3 & 4 \\ 4 & 2 & 1 & 2 & 3 & 2 & 1 \\ 5 & 1 & 2 & 3 & 4 & 1 & 2 \end{bmatrix}.$$

(b)

L	D(2)	D(3)	D(4)	D(5)	D(6)	D(7)
∅	∞	∞	∞	1	∞	∞
{5}	∞	∞	2	1	∞	∞
{5, 4}	∞	3	2	1	∞	∞
{5, 4, 3}	∞	3	2	1	4	∞
{5, 4, 3, 6}	∞	3	2	1	4	5
{5, 4, 3, 6, 7}	6	3	2	1	4	5

6. (a)

$$W_0 = W_1 = \begin{bmatrix} \infty & 1 & 2 & \infty & \infty & \infty \\ \infty & \infty & \infty & 2 & 3 & \infty \\ \infty & \infty & \infty & 5 & \infty & \infty \\ \infty & \infty & \infty & \infty & \infty & 2 \\ \infty & \infty & \infty & \infty & \infty & 4 \\ \infty & \infty & \infty & \infty & \infty & \infty \end{bmatrix}, \qquad W_2 = W_3 = \begin{bmatrix} \infty & 1 & 2 & 3 & 4 & \infty \\ \infty & \infty & \infty & 2 & 3 & \infty \\ \infty & \infty & \infty & 5 & \infty & \infty \\ \infty & \infty & \infty & \infty & \infty & 2 \\ \infty & \infty & \infty & \infty & \infty & 4 \\ \infty & \infty & \infty & \infty & \infty & \infty \end{bmatrix},$$

$$W_4 = W_5 = W_6 = \begin{bmatrix} \infty & 1 & 2 & 3 & 4 & 5 \\ \infty & \infty & \infty & 2 & 3 & 4 \\ \infty & \infty & \infty & 5 & \infty & 7 \\ \infty & \infty & \infty & \infty & \infty & 2 \\ \infty & \infty & \infty & \infty & \infty & 4 \\ \infty & \infty & \infty & \infty & \infty & \infty \end{bmatrix}.$$

(b)

$$W_0 = W_1 = \begin{bmatrix} -\infty & 1 & 2 & -\infty & -\infty & -\infty \\ -\infty & -\infty & -\infty & 2 & 3 & -\infty \\ -\infty & -\infty & -\infty & 5 & -\infty & -\infty \\ -\infty & -\infty & -\infty & -\infty & -\infty & 2 \\ -\infty & -\infty & -\infty & -\infty & -\infty & 4 \\ -\infty & -\infty & -\infty & -\infty & -\infty & -\infty \end{bmatrix}, \quad W_2 = \begin{bmatrix} -\infty & 1 & 2 & 3 & 4 & -\infty \\ -\infty & -\infty & -\infty & 2 & 3 & -\infty \\ -\infty & -\infty & -\infty & 5 & -\infty & -\infty \\ -\infty & -\infty & -\infty & -\infty & -\infty & 2 \\ -\infty & -\infty & -\infty & -\infty & -\infty & 4 \\ -\infty & -\infty & -\infty & -\infty & -\infty & -\infty \end{bmatrix},$$

$$W_3 = \begin{bmatrix} -\infty & 1 & 2 & 7 & 4 & -\infty \\ -\infty & -\infty & -\infty & 2 & 3 & -\infty \\ -\infty & -\infty & -\infty & 5 & -\infty & -\infty \\ -\infty & -\infty & -\infty & -\infty & -\infty & 2 \\ -\infty & -\infty & -\infty & -\infty & -\infty & 4 \\ -\infty & -\infty & -\infty & -\infty & -\infty & -\infty \end{bmatrix}, \quad W_4 = \begin{bmatrix} -\infty & 1 & 2 & 7 & 4 & 9 \\ -\infty & -\infty & -\infty & 2 & 3 & 4 \\ -\infty & -\infty & -\infty & 5 & -\infty & 7 \\ -\infty & -\infty & -\infty & -\infty & -\infty & 2 \\ -\infty & -\infty & -\infty & -\infty & -\infty & 4 \\ -\infty & -\infty & -\infty & -\infty & -\infty & -\infty \end{bmatrix},$$

$$W_5 = W_6 = \begin{bmatrix} -\infty & 1 & 2 & 7 & 4 & 9 \\ -\infty & -\infty & -\infty & 2 & 3 & 7 \\ -\infty & -\infty & -\infty & 5 & -\infty & 7 \\ -\infty & -\infty & -\infty & -\infty & -\infty & 2 \\ -\infty & -\infty & -\infty & -\infty & -\infty & 4 \\ -\infty & -\infty & -\infty & -\infty & -\infty & -\infty \end{bmatrix}.$$

7.

$$W^* = \begin{bmatrix} \infty & 8 & 7 & 5 & 2 \\ \infty & \infty & \infty & \infty & \infty \\ \infty & 1 & \infty & \infty & \infty \\ \infty & 3 & 2 & \infty & \infty \\ \infty & 6 & 5 & 3 & \infty \end{bmatrix}.$$

8.

L	D(1)	D(2)	D(4)	D(5)	D(6)	D(7)
$\emptyset$	∞	∞	1	5	7	∞
$\{4\}$	∞	∞	1	4	7	∞
$\{4, 5\}$	∞	∞	1	4	5	7

no more changes.

9. (a) $\quad D_0 = D_1 = [\ -\infty\ \ 1\ \ 2\ -\infty\ -\infty\ -\infty\ -\infty\ ]$

$\qquad\qquad D_2 = [\ -\infty\ \ 1\ \ 2\ \ 5\ \ 4\ -\infty\ -\infty\ ]$

$\qquad\qquad D_3 = [\ -\infty\ \ 1\ \ 2\ \ 5\ \ 7\ \ 9\ -\infty\ ]$

$\qquad\qquad D_4 = [\ -\infty\ \ 1\ \ 2\ \ 5\ \ 8\ \ 13\ -\infty\ ]$

$\qquad\qquad D_5 = [\ -\infty\ \ 1\ \ 2\ \ 5\ \ 8\ \ 13\ \ 11\ ]$

$\qquad\qquad D_6 = [\ -\infty\ \ 1\ \ 2\ \ 5\ \ 8\ \ 13\ \ 16\ ]$

(b) The only allowable relabeling is $s = v_1$, $a = v_2$, $c = v_3$, $b = v_4$, $d = v_5$.
$\quad D_0 = D_1 = [\ -\infty\ \ 2\ -\infty\ \ 6\ -\infty\ ]$, $D_2 = [\ -\infty\ \ 2\ \ 4\ \ 6\ \ 3\ ]$,
$\quad D_3 = [\ -\infty\ \ 2\ \ 4\ \ 7\ \ 3\ ]$, $D_4 = [\ -\infty\ \ 2\ \ 4\ \ 7\ \ 8\ ]$.

10. (a) DIJKSTRA'S algorithm gives

L	D(2)	D(3)	D(4)
$\emptyset$	4	2	∞
$\{3\}$	4	2	6
$\{3, 2\}$	4	2	5

when in fact $W^*(v_1, v_3) = 4 - 4 = 0$ and $W^*(v_1, v_4) = 0 + 4 = 4$.

(b) No. The algorithm would yield $W^*(v_1, v_3) = 0$ correctly, but still give $D(4) = 5$, rather than 4.

11. (a) The algorithm would give

L	D(2)	D(3)	D(4)
Ø	5	6	-∞
{3}	5	6	9
		no change	

whereas $M(1,3) = 9$ and $M(1,4) = 12$.

(b) The algorithm would give

L	D(2)	D(3)	D(4)
Ø	6	4	-∞
{3}	6	4	5
		no change	

whereas $M(1,3) = 10$ and $M(1,4) = 11$.

(c) Both algorithms would still fail to give correct values of $M(1,4)$.

12. (a) We claim that MAX-WEIGHT and the max-modified WARSHALL'S algorithm produce the same final value of $W[1,j]$, namely $M(1,j)$, for each j.

Consider the max-modified WARSHALL'S algorithm applied to a digraph with a reverse sorted labeling. The algorithm guarantees that for every $k \geq 1$ and every $j \geq 1$

$$W_k[1,j] = \max\{W_{k-1}[1,j], W_{k-1}[1,k] + W_{k-1}[k,j]\}.$$

The proof that WARSHALL'S algorithm works shows that $W_{k-1}[k,j]$ is the largest weight of a path from k to j with intermediate vertices in $\{1, \ldots, k-1\}$. Since the digraph is reverse sorted, no path starting at k goes through any of $1, \ldots, k-1$, and so $W_{k-1}[k,j]$

$= W(k,j)$. Thus for every j

(*) $W_k[1,j] = \max\{W_{k-1}[1,j], W_{k-1}[1,k] + W(k,j)\}.$

Assume inductively that $W_{k-1}[1,j]$ and $W_{k-1}[1,k]$ are the values of $W[1,j]$ and $W[1,k]$ produced by MAX-WEIGHT at the end of the loop for k - 1. [Here it is convenient to include a loop for $k = 1$ in MAX-WEIGHT, even though nothing happens then because $W_0[1,1] = -\infty$.] Then (*) says that for every j $W_k[1,j]$ is the value of $W[1,j]$ produced by MAX-WEIGHT at the end of the loop for k. Thus, by induction, MAX-WEIGHT and the max-modified WARSHALL'S algorithm produce the same values of $W[1,j]$ at each stage, and hence give the same end results.

[In passing, we note that the k-loop in MAX-WEIGHT could just as well have been placed inside the j-loop.]

137

(b) The comparison and replacement steps take a fixed amount of time, and the j-loop repeats at most $n - 2$ times for each of the $n - 2$ passes through the k-loop.

Section 8.4

1. (a)

$$P_0 = \begin{bmatrix} 0 & 2 & 3 & 0 & 0 & 0 & 0 \\ 0 & 0 & 0 & 4 & 5 & 0 & 0 \\ 0 & 0 & 0 & 4 & 5 & 6 & 0 \\ 0 & 0 & 0 & 0 & 5 & 6 & 0 \\ 0 & 0 & 0 & 0 & 0 & 6 & 7 \\ 0 & 0 & 0 & 0 & 0 & 0 & 7 \\ 0 & 0 & 0 & 0 & 0 & 0 & 0 \end{bmatrix}, \quad P_{final} = \begin{bmatrix} 0 & 2 & 3 & 3 & 2 & 2 & 2 \\ 0 & 0 & 0 & 4 & 5 & 5 & 5 \\ 0 & 0 & 0 & 4 & 4 & 4 & 4 \\ 0 & 0 & 0 & 0 & 5 & 5 & 5 \\ 0 & 0 & 0 & 0 & 0 & 6 & 7 \\ 0 & 0 & 0 & 0 & 0 & 0 & 7 \\ 0 & 0 & 0 & 0 & 0 & 0 & 0 \end{bmatrix}.$$

(b) P_0 as in (a).

$$P_{final} = \begin{bmatrix} 0 & 2 & 3 & 2 & 2 & 2 & 2 \\ 0 & 0 & 0 & 4 & 4 & 4 & 4 \\ 0 & 0 & 0 & 4 & 5 & 4 & 4 \\ 0 & 0 & 0 & 0 & 5 & 6 & 6 \\ 0 & 0 & 0 & 0 & 0 & 6 & 6 \\ 0 & 0 & 0 & 0 & 0 & 0 & 7 \\ 0 & 0 & 0 & 0 & 0 & 0 & 0 \end{bmatrix}.$$

2.

$$W_0 = \begin{bmatrix} 0 & 8 & \infty & \infty & 1 \\ 8 & 0 & 3 & \infty & \infty \\ \infty & 3 & 0 & 1 & \infty \\ \infty & \infty & 1 & 0 & 2 \\ 1 & \infty & \infty & 2 & 0 \end{bmatrix}, \quad P_0 = \begin{bmatrix} 0 & 2 & 1 & 1 & 5 \\ 1 & 0 & 3 & 2 & 2 \\ 3 & 2 & 0 & 4 & 3 \\ 4 & 4 & 3 & 0 & 5 \\ 1 & 5 & 5 & 4 & 0 \end{bmatrix},$$

$$W_1 = \begin{bmatrix} 0 & 8 & \infty & \infty & 1 \\ 8 & 0 & 3 & \infty & 9 \\ \infty & 3 & 0 & 1 & \infty \\ \infty & \infty & 1 & 0 & 2 \\ 1 & 9 & \infty & 2 & 0 \end{bmatrix}, \quad P_1 = \begin{bmatrix} 0 & 2 & 1 & 1 & 5 \\ 1 & 0 & 3 & 2 & 1 \\ 3 & 2 & 0 & 4 & 3 \\ 4 & 4 & 3 & 0 & 5 \\ 1 & 1 & 5 & 4 & 0 \end{bmatrix},$$

$$W_2 = \begin{bmatrix} 0 & 8 & 11 & \infty & 1 \\ 8 & 0 & 3 & \infty & 9 \\ 11 & 3 & 0 & 1 & 12 \\ \infty & \infty & 1 & 0 & 2 \\ 1 & 9 & 12 & 2 & 0 \end{bmatrix}, \quad P_2 = \begin{bmatrix} 0 & 2 & 2 & 1 & 5 \\ 1 & 0 & 3 & 2 & 1 \\ 2 & 2 & 0 & 4 & 2 \\ 4 & 4 & 3 & 0 & 5 \\ 1 & 1 & 1 & 4 & 0 \end{bmatrix},$$

$$W_3 = \begin{bmatrix} 0 & 8 & 11 & 12 & 1 \\ 8 & 0 & 3 & 4 & 9 \\ 11 & 3 & 0 & 1 & 12 \\ 12 & 4 & 1 & 0 & 2 \\ 1 & 9 & 12 & 2 & 0 \end{bmatrix}, \quad P_3 = \begin{bmatrix} 0 & 2 & 2 & 2 & 5 \\ 1 & 0 & 3 & 3 & 1 \\ 2 & 2 & 0 & 4 & 2 \\ 3 & 3 & 3 & 0 & 5 \\ 1 & 1 & 1 & 4 & 0 \end{bmatrix},$$

$$W_4 = \begin{bmatrix} 0 & 8 & 11 & 12 & 1 \\ 8 & 0 & 3 & 4 & 6 \\ 11 & 3 & 0 & 1 & 3 \\ 12 & 4 & 1 & 0 & 2 \\ 1 & 6 & 3 & 2 & 0 \end{bmatrix}, \qquad P_4 = \begin{bmatrix} 0 & 2 & 2 & 2 & 5 \\ 1 & 0 & 3 & 3 & 3 \\ 2 & 2 & 0 & 4 & 4 \\ 3 & 3 & 3 & 0 & 5 \\ 1 & 4 & 4 & 4 & 0 \end{bmatrix},$$

$$W^* = W_5 = \begin{bmatrix} 0 & 7 & 4 & 3 & 1 \\ 7 & 0 & 3 & 4 & 6 \\ 4 & 3 & 0 & 1 & 3 \\ 3 & 4 & 1 & 0 & 2 \\ 1 & 6 & 3 & 2 & 0 \end{bmatrix}, \qquad P^* = P_5 = \begin{bmatrix} 0 & 5 & 5 & 5 & 5 \\ 3 & 0 & 3 & 3 & 3 \\ 4 & 2 & 0 & 4 & 4 \\ 5 & 3 & 3 & 0 & 5 \\ 1 & 4 & 4 & 4 & 0 \end{bmatrix}.$$

3. (a)

k	D(2)	D(3)	D(4)	D(5)	P(2)	P(3)	P(4)	P(5)
1	2	1*	7	∞	1	1	1	0
3	2*	1	7	4	1	1	1	3
2	2	1	6	4*	1	1	2	3
5	2	1	5*	4	1	1	5	3

Asterisk (*) in column D(k) marks the time that k is chosen for L and D(k) is frozen.

(b)

k	D(2)	D(3)	D(4)	D(5)	P(2)	P(3)	P(4)	P(5)
1	∞	4*	∞	∞	0	1	0	0
3	∞	4	6*	∞	0	1	3	0
4	∞	4	6	9*	0	1	3	4
5	∞	4	6	9	0	1	3	4

* in column D(k) marks the time that k is chosen for L and D(k) is frozen.

4. The min-paths are: $v_1 v_2$; $v_1 v_2 v_5 v_4$; $v_1 v_2 v_5 v_4 v_6$.

5. (a)

$$W_0 = W_1 = \begin{bmatrix} \infty & 1 & \infty & 7 & \infty \\ \infty & \infty & 4 & 2 & \infty \\ \infty & \infty & \infty & \infty & 3 \\ \infty & \infty & 1 & \infty & 5 \\ \infty & \infty & \infty & \infty & \infty \end{bmatrix}, \qquad P_0 = P_1 = \begin{bmatrix} 0 & 2 & 0 & 4 & 0 \\ 0 & 0 & 3 & 4 & 0 \\ 0 & 0 & 0 & 0 & 5 \\ 0 & 0 & 3 & 0 & 5 \\ 0 & 0 & 0 & 0 & 0 \end{bmatrix},$$

$$W_2 = \begin{bmatrix} \infty & 1 & 5 & 3 & \infty \\ \infty & \infty & 4 & 2 & \infty \\ \infty & \infty & \infty & \infty & 3 \\ \infty & \infty & 1 & \infty & 5 \\ \infty & \infty & \infty & \infty & \infty \end{bmatrix}, \qquad P_2 = \begin{bmatrix} 0 & 2 & 2 & 2 & 0 \\ 0 & 0 & 3 & 4 & 0 \\ 0 & 0 & 0 & 0 & 5 \\ 0 & 0 & 3 & 0 & 5 \\ 0 & 0 & 0 & 0 & 0 \end{bmatrix},$$

$$W_3 = \begin{bmatrix} \infty & 1 & 5 & 3 & 8 \\ \infty & \infty & 4 & 2 & 7 \\ \infty & \infty & \infty & \infty & 3 \\ \infty & \infty & 1 & \infty & 4 \\ \infty & \infty & \infty & \infty & \infty \end{bmatrix}, \qquad P_3 = \begin{bmatrix} 0 & 2 & 2 & 2 & 2 \\ 0 & 0 & 3 & 4 & 3 \\ 0 & 0 & 0 & 0 & 5 \\ 0 & 0 & 3 & 0 & 3 \\ 0 & 0 & 0 & 0 & 0 \end{bmatrix},$$

$$W_4 = \begin{bmatrix} \infty & 1 & 4 & 3 & 7 \\ \infty & \infty & 3 & 2 & 6 \\ \infty & \infty & \infty & \infty & 3 \\ \infty & \infty & 1 & \infty & 4 \\ \infty & \infty & \infty & \infty & \infty \end{bmatrix}, \qquad P_4 = \begin{bmatrix} 0 & 2 & 2 & 2 & 2 \\ 0 & 0 & 4 & 4 & 4 \\ 0 & 0 & 0 & 0 & 5 \\ 0 & 0 & 3 & 0 & 3 \\ 0 & 0 & 0 & 0 & 0 \end{bmatrix}.$$

Also, $W_5 = W^* = W_4$ and $P_5 = P^* = P_4$.

(b)

$$W_0 = W_1 = \begin{bmatrix} -\infty & 1 & -\infty & 7 & -\infty \\ -\infty & -\infty & 4 & 2 & -\infty \\ -\infty & -\infty & -\infty & -\infty & 3 \\ -\infty & -\infty & 1 & -\infty & 5 \\ -\infty & -\infty & -\infty & -\infty & -\infty \end{bmatrix}, \qquad P_0 = P_1 = \begin{bmatrix} 0 & 2 & 0 & 4 & 0 \\ 0 & 0 & 3 & 4 & 0 \\ 0 & 0 & 0 & 0 & 5 \\ 0 & 0 & 3 & 0 & 5 \\ 0 & 0 & 0 & 0 & 0 \end{bmatrix},$$

$$W_2 = \begin{bmatrix} -\infty & 1 & 5 & 7 & -\infty \\ -\infty & -\infty & 4 & 2 & 7 \\ -\infty & -\infty & -\infty & -\infty & 3 \\ -\infty & -\infty & 1 & -\infty & 5 \\ -\infty & -\infty & -\infty & -\infty & -\infty \end{bmatrix}, \qquad P_2 = \begin{bmatrix} 0 & 2 & 2 & 4 & 0 \\ 0 & 0 & 3 & 4 & 0 \\ 0 & 0 & 0 & 0 & 5 \\ 0 & 0 & 3 & 0 & 5 \\ 0 & 0 & 0 & 0 & 0 \end{bmatrix},$$

$$W_3 = \begin{bmatrix} -\infty & 1 & 5 & 7 & 8 \\ -\infty & -\infty & 4 & 2 & 7 \\ -\infty & -\infty & -\infty & -\infty & 3 \\ -\infty & -\infty & 1 & -\infty & 5 \\ -\infty & -\infty & -\infty & -\infty & -\infty \end{bmatrix}, \qquad P_3 = \begin{bmatrix} 0 & 2 & 2 & 4 & 2 \\ 0 & 0 & 3 & 4 & 3 \\ 0 & 0 & 0 & 0 & 5 \\ 0 & 0 & 3 & 0 & 5 \\ 0 & 0 & 0 & 0 & 0 \end{bmatrix},$$

$$W_4 = W_5 = M = \begin{bmatrix} -\infty & 1 & 8 & 7 & 12 \\ -\infty & -\infty & 4 & 2 & 7 \\ -\infty & -\infty & -\infty & -\infty & 3 \\ -\infty & -\infty & 1 & -\infty & 5 \\ -\infty & -\infty & -\infty & -\infty & -\infty \end{bmatrix}, \qquad P_4 = P_5 = \begin{bmatrix} 0 & 2 & 4 & 4 & 4 \\ 0 & 0 & 3 & 4 & 3 \\ 0 & 0 & 0 & 0 & 5 \\ 0 & 0 & 3 & 0 & 5 \\ 0 & 0 & 0 & 0 & 0 \end{bmatrix}.$$

6. (a)

$$W_0 = \begin{bmatrix} \infty & \infty & 1 & \infty & 1 & \infty \\ 1 & \infty & 1 & 1 & \infty & \infty \\ \infty & \infty & \infty & \infty & \infty & \infty \\ \infty & \infty & 1 & \infty & 1 & 1 \\ \infty & \infty & 1 & \infty & \infty & \infty \\ 1 & 1 & \infty & \infty & 1 & \infty \end{bmatrix}, \qquad W_1 = \begin{bmatrix} \infty & \infty & 1 & \infty & 1 & \infty \\ 1 & \infty & 1 & 1 & 2 & \infty \\ \infty & \infty & \infty & \infty & \infty & \infty \\ \infty & \infty & 1 & \infty & 1 & 1 \\ \infty & \infty & 1 & \infty & \infty & \infty \\ 1 & 1 & 2 & \infty & 1 & \infty \end{bmatrix},$$

$$W_2 = W_3 = \begin{bmatrix} \infty & \infty & 1 & \infty & 1 & \infty \\ 1 & \infty & 1 & 1 & 2 & \infty \\ \infty & \infty & \infty & \infty & \infty & \infty \\ \infty & \infty & 1 & \infty & 1 & 1 \\ \infty & \infty & 1 & \infty & \infty & \infty \\ 1 & 1 & 2 & 2 & 1 & \infty \end{bmatrix}, \qquad W_4 = W_5 = \begin{bmatrix} \infty & \infty & 1 & \infty & 1 & \infty \\ 1 & \infty & 1 & 1 & 2 & 2 \\ \infty & \infty & \infty & \infty & \infty & \infty \\ \infty & \infty & 1 & \infty & 1 & 1 \\ \infty & \infty & 1 & \infty & \infty & \infty \\ 1 & 1 & 2 & 2 & 1 & 3 \end{bmatrix},$$

$$W_6 = W^* = \begin{bmatrix} \infty & \infty & 1 & \infty & 1 & \infty \\ 1 & 3 & 1 & 1 & 2 & 2 \\ \infty & \infty & \infty & \infty & \infty & \infty \\ 2 & 2 & 1 & 3 & 1 & 1 \\ \infty & \infty & 1 & \infty & \infty & \infty \\ 1 & 1 & 2 & 2 & 1 & 3 \end{bmatrix}, \quad \text{so } M_R = \begin{bmatrix} 0 & 0 & 1 & 0 & 1 & 0 \\ 1 & 1 & 1 & 1 & 1 & 1 \\ 0 & 0 & 0 & 0 & 0 & 0 \\ 1 & 1 & 1 & 1 & 1 & 1 \\ 0 & 0 & 1 & 0 & 0 & 0 \\ 1 & 1 & 1 & 1 & 1 & 1 \end{bmatrix}.$$

(b) No; $v_2 v_4 v_6 v_2$ is a cycle. If the digraph were acyclic, all diagonal entries in M_R would be 0.

7. (a) Create a row matrix P, with $P[j] = 1$ initially if there is an edge from v_1 to v_j and $P[j] = 0$ otherwise. Add the line

replace $P[j]$ by k.

(b) Part of this exercise is solved in Example 3 of § 8.3. The sequence of row matrices P_k is as follows.
$P_0 = P_1 = [0 \ 1 \ 1 \ 0 \ 0 \ 0]$, $P_2 = [0 \ 1 \ 1 \ 2 \ 2 \ 0]$, $P_3 = [0 \ 1 \ 1 \ 3 \ 2 \ 0]$,
$P_4 = P_5 = [0 \ 1 \ 1 \ 3 \ 2 \ 4]$.

8. (a)

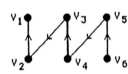

(b) Yes.
(c)

$$M_R = \begin{bmatrix} 0 & 0 & 0 & 0 & 0 & 0 \\ 1 & 0 & 0 & 0 & 0 & 0 \\ 1 & 1 & 0 & 0 & 0 & 0 \\ 1 & 1 & 1 & 0 & 0 & 0 \\ 1 & 1 & 1 & 1 & 0 & 0 \\ 1 & 1 & 1 & 1 & 1 & 0 \end{bmatrix}.$$

9. (a)

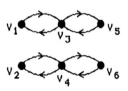

(b) No; for instance $v_1 v_3 v_1$ is a cycle.

(c)

$$M_R = \begin{bmatrix} 1 & 0 & 1 & 0 & 1 & 0 \\ 0 & 1 & 0 & 1 & 0 & 1 \\ 1 & 0 & 1 & 0 & 1 & 0 \\ 0 & 1 & 0 & 1 & 0 & 1 \\ 1 & 0 & 1 & 0 & 1 & 0 \\ 0 & 1 & 0 & 1 & 0 & 1 \end{bmatrix}.$$

10. Initially, $W[i,j] = 1$ if there is an edge from v_i to v_j, and $W[i,j] = 0$ otherwise. Change the update step to

   if $W[i,j] < \min\{W[i,k], W[k,j]\}$, then
      replace $W[i,j]$ by $\min\{W[i,k], W[k,j]\}$

or to

   $W[i,j] := \max\{W[i,j], \min\{W[i,k], W[k,j]\}\}$.

11. (a) Initialization of P takes time $O(n)$, and the replacement step still only takes constant time during each pass through the loop.

   (b) $O(n^3)$. The number of inside loops remains n^3, and the time to update W and P is constant for each loop.

CHAPTER 9

This new chapter presents basic material on probability: independence, random variables and some important distributions. Though we expect our readers to be thinking of finite sample spaces and discrete random variables, the main facts are still true, suitably interpreted, in a more general setting.

Section 9.1 introduces conditional probability and independence. Our point of view is that P(| S) is a probability on Ω determined by the probability P( ) and the choice of S. Independence is of course a fundamental concept. Give examples of events that are independent and ones that aren't. Illustrate Bayes' Formula with a couple of examples. Exercise 19, drawn from actual facts, is a real surprise to most people.

In § 9.2 we move away from sample spaces toward random variables and begin to think of density and distribution functions. As you work examples, point out that the sample spaces are largely irrelevant to answering the questions we care about.

Section 9.3 gives the basic facts about mean and variance. Illustrate the theorems with examples, and leave the proofs for the students to read. This section has enough facts in it to warrant two class periods.

The last section of the chapter discusses binomial and geometric distributions in some detail and then, with a nod toward the central limit theorem, shows how to use the normal distribution to get approximate answers for binomial distributions. [To avoid further complications, we have ignored the continuity correction that would give better approximations, especially in Example 9 and Exercises 6 and 7.] Give a variety of examples. Point out especially how the mean and standard deviation change with increasing n in Theorem 1. Theorem 2 is the link to the normal distribution; its proof can be left to the reader.

We would be especially interested to hear whether this new chapter is pitched at the right level for your students, since we haven't had a chance to try it yet on ours.

Section 9.1

1. (a) $\dfrac{\binom{3}{2}}{\binom{11}{2}} = \dfrac{3}{55}$.   (b) $\dfrac{\binom{8}{2}}{\binom{11}{2}} = \dfrac{28}{55}$.   (c) $\dfrac{3 \cdot 8}{\binom{11}{2}} = \dfrac{24}{55}$.

2. (a) $\frac{9}{121}$. (b) $\frac{64}{121}$. (c) $\frac{48}{121}$.

With replacement, it's easier to get two marbles of the same color.

3. (a) $P(B_0) = P(R_0) = P(E) = \frac{1}{2}$ and $P(B_0 \cap R_0') = P(B_0 \cap E) = P(R_0 \cap E) = \frac{1}{4}$.

(b) $P(B_0 \mid E \cap R_0) = 1 \neq P(B_0)$.

4. See answer to Exercise 5.

5. $\{S, L\}$ are dependent since $P(S \cap L) = \frac{6}{36}$ while $P(S) \cdot P(L) = \frac{15}{36} \cdot \frac{15}{36}$. $\{S, E\}$ are dependent since $P(S \cap E) = \frac{3}{36}$ while $P(S) \cdot P(E) = \frac{15}{36} \cdot \frac{6}{36}$. $\{L, E\}$ are dependent since $P(L \mid E) = 0 \neq P(L)$. Similarly for $\{L, G\}$.

6. (a) $\frac{P(\text{red} \geq 5 \text{ and sum} = 9)}{P(\text{sum} = 9)} = \frac{1}{2}$. (b) $\frac{P(\text{red} \geq 5 \text{ and sum} \geq 9)}{P(\text{sum} \geq 9)} = \frac{7}{10}$.

7. No. $P(B \mid A) = \frac{1}{2}$ while $P(B) = \frac{\binom{4}{2}}{2^4} = \frac{3}{8}$.

8. (a) $\frac{1}{2}$. (b) $\frac{8}{11}$.

9. (a) .25. (b) .7. (c) No, $P(A \mid B) = .25 \neq P(A)$.

(d) No, $P(A^c \cap B) = .3 \neq P(A^c) \cdot P(B) = .28$.

10. (a) $P(A \mid B) = P(A) = .4$.

(b) $P(A \cup B) = P(A) + P(B) - P(A \cap B) = .4 + .6 - .24 = .76$.

(c) $P(B) - P(A \cap B) = .6 - .24 = .36$.

11. (a) $\frac{4}{52} \cdot \frac{3}{51} \cdot \frac{2}{50} \approx .00018$. (b) $\frac{4}{52} \cdot \frac{4}{51} \cdot \frac{4}{50} \approx .00048$.

(c) $1 - \frac{48}{52} \cdot \frac{47}{51} \cdot \frac{46}{50} \approx .217$.

12. Half of the 5108 flushes involve red cards only. There are $\binom{26}{5} = 65,780$ five card hands with all cards red. So the answer is $\frac{2554}{65,780} \approx .0388$.

13. (a) $P(B) = \frac{1}{3} \cdot \frac{2}{3} + \frac{1}{3} \cdot \frac{2}{5} + \frac{1}{3} \cdot \frac{1}{2} = \frac{47}{90}$.

(b) $P(U_1 \mid B) = \dfrac{\frac{1}{3} \cdot \frac{2}{3}}{P(B)} = \dfrac{20}{47}$. $\quad P(U_2 \mid B) = \dfrac{\frac{1}{3} \cdot \frac{2}{5}}{P(B)} = \dfrac{12}{47}$. $\quad P(U_3 \mid B) = \dfrac{\frac{1}{3} \cdot \frac{1}{2}}{P(B)} = \dfrac{15}{47}$.

(c) $P(B \cap U_1) = \dfrac{1}{3} \cdot \dfrac{2}{3} = \dfrac{2}{9}$.

14. (a) $0 + \dfrac{1}{3} \cdot \dfrac{\binom{3}{2}}{\binom{5}{2}} + \dfrac{1}{3} \cdot \dfrac{1}{\binom{4}{2}} = \dfrac{7}{45}$. $\qquad$ (b) $\dfrac{1}{3} \cdot \dfrac{1}{\binom{3}{2}} + \dfrac{1}{3} \cdot \dfrac{1}{\binom{5}{2}} + \dfrac{1}{3} \cdot \dfrac{1}{\binom{4}{2}} = \dfrac{1}{5}$.

15. (a) $\dfrac{5}{9}$. $\qquad$ (b) 0.

16. (a) $P(\text{all red}) = \dfrac{1}{3} \cdot \dfrac{3}{5} \cdot \dfrac{2}{4} = \dfrac{1}{10}$; $\quad P(\text{all black}) = \dfrac{2}{3} \cdot \dfrac{2}{5} \cdot \dfrac{2}{4} = \dfrac{2}{15}$.

17. (a) $P(C) = P(E) \cdot P(C \mid E) + P(F) \cdot P(C \mid F) + P(G) \cdot P(C \mid G) = .043$.

(b) $P(E \mid C) = \dfrac{10}{43} \approx .23$, $\quad P(F \mid C) = \dfrac{21}{43} \approx .49$, $\quad P(G \mid C) = \dfrac{12}{43} \approx .28$.

18. (a) $\dfrac{3}{4}$. $\qquad$ (b) $\dfrac{2}{3}$. $\qquad$ (c) $\dfrac{5}{8}$.

19. (a) $P(D) = P(N^c \cap D) + P(N \cap D) = .0041$, so $P(N^c \mid D) = \dfrac{P(N^c \cap D)}{P(D)} = \dfrac{.004}{.0041} \approx .9756$.

(b) Since $P(N^c) = .044$, $P(N) = .956$ and so $P(N \cap D^c) = P(N) - P(N \cap D) = .9559$.
Hence $P((N^c \cap D) \cup (N \cap D^c)) = .004 + .9559 = .9599$. This is the probability that the test confirms the subject's condition.

(c) $P(D \mid N^c) = \dfrac{P(D \cap N^c)}{P(N^c)} = \dfrac{.004}{.044} \approx .091$. Thus the probability of having the disease, given a positive test is less than .10. The following table may help clarify the situation.

	D [diseased]	D^c [not diseased]
N^c [tests positive]	.004	.04
N [tests negative]	.0001	.9559

20. $P(A \cap B^c) = P(A) - P(A \cap B) = P(A) - P(A) \cdot P(B) = P(A) \cdot [1 - P(B)] = P(A) \cdot P(B^c)$.
Similarly $P(A^c \cap B) = P(A^c) \cdot P(B)$ and $P(A^c \cap B^c) = P(A^c) \cdot P(B^c)$.

21. (a) $1 - (1 - q)^n$. We are assuming that the failures of the components are independent.

(b) $1 - (.99)^{100} \approx .634$. This is close to $1 - \dfrac{1}{e} \approx .632$ because $\lim_n (1 - \tfrac{1}{n})^n = \dfrac{1}{e}$.

(c) $1 - (.999)^{100} \approx .0952$.

(d) $1 - (.9)^{100} \approx .99997$.

22. $P(A \mid B) > P(A) \iff P(A \cap B) > P(A) \cdot P(B) \iff P(B \mid A) > P(B).$

23. (a) $\dfrac{\binom{18}{5}}{\binom{20}{5}} = \dfrac{15 \cdot 14}{20 \cdot 19} = \dfrac{21}{38} \approx .55.$ (b) $\dfrac{\binom{18}{10}}{\binom{20}{10}} = \dfrac{9}{38} \approx .24.$

24. Since $P(A \cap B) = P(A) + P(B) - P(A \cup B) = \frac{4}{3} - P(A \cup B) \geq \frac{4}{3} - 1 = \frac{1}{3}$, we have

$P(A \mid B) = \dfrac{P(A \cap B)}{P(B)} = \dfrac{3}{2} \cdot P(A \cap B) \geq \dfrac{3}{2} \cdot \dfrac{1}{3} = \dfrac{1}{2}$.

25. No. For example, toss a fair coin three times and let A_k = "k th toss is a head." Then A_1, A_2, A_3 are independent, but $A_1 \cap A_2$ and $A_1 \cap A_3$ are not. Indeed

$$P(A_1 \cap A_3 \mid A_1 \cap A_2) = \frac{1}{2} \neq \frac{1}{4} = P(A_1 \cap A_3).$$

26. Use induction on n. Given events $A_1, A_2, \ldots, A_{n+1}$, apply the induction hypothesis to $B_1 = A_1 \cap A_2$, $B_2 = A_3, \ldots,$ $B_n = A_{n+1}$. Then

$P(A_1 \cap A_2 \cap \cdots \cap A_n \cap A_{n+1})$
$\quad = P(A_1 \cap A_2) \cdot P(A_3 \mid A_1 \cap A_2) \cdots P(A_{n+1} \mid A_1 \cap A_2 \cap \cdots \cap A_n)$
$\quad = P(A_1) \cdot P(A_2 \mid A_1) P(A_3 \mid A_1 \cap A_2) \cdots P(A_{n+1} \mid A_1 \cap A_2 \cap \cdots \cap A_n).$

For $n = 3$, we have
$P(A_1 \cap A_2 \cap A_3) = P(A_1 \cap A_2) \cdot P(A_3 \mid A_1 \cap A_2) = P(A_1) \cdot P(A_2 \mid A_1) \cdot P(A_3 \mid A_1 \cap A_2).$
Then for $n = 4$, we get
$P(A_1 \cap A_2 \cap A_3 \cap A_4) = P(A_1 \cap A_2) \cdot P(A_3 \mid A_1 \cap A_2) \cdot P(A_4 \mid A_1 \cap A_2 \cap A_3) =$
$\quad P(A_1) \cdot P(A_2 \mid A_1) \cdot P(A_3 \mid A_1 \cap A_2) \cdot P(A_4 \mid A_1 \cap A_2 \cap A_3).$

27. (a) No. If true, then A and B independent and B and A independent would imply that A and A are independent, which generally fails by part (b).
(b) No. Only if $P(A) = 0$ or $P(A) = 1$, since $P(A \cap A) = P(A) \cdot P(A)$ implies $P(A)^2 = P(A)$.
(c) Absolutely not, unless $P(A) = 0$ or $P(B) = 0$.

28. (a) $P(B \mid A) = \dfrac{P(B \cap A)}{P(A)} > P(A \cap B) = P(B) \cdot \dfrac{P(A \cap B)}{P(B)} = P(B) \cdot P(A \mid B) = P(B).$
(b) If $P(A) = 1$, then $P(A \cup B) = 1$ and so $P(A \cap B) = P(A) + P(B) - P(A \cup B) = P(B)$. Hence $P(A \mid B) = 1$ automatically and also $P(B \mid A) = P(B)$.

29. (a) $\dfrac{P^*(E)}{P^*(F)} = \dfrac{P(E \mid S)}{P(F \mid S)} = \dfrac{\dfrac{P(E \cap S)}{P(S)}}{\dfrac{P(F \cap S)}{P(S)}} = \dfrac{P(E \cap S)}{P(F \cap S)}.$

(b) Set $F = \Omega$ in part (a).

Section 9.2

1. (a) $\{3, 4, 5, 6, \ldots , 18\}$. (b) $\{n, n + 1, n + 2, \ldots , 6n\}$.

2. (a) $\dfrac{21}{36} = \dfrac{7}{12}$. (b) $\dfrac{27}{36} = \dfrac{3}{4}$. (c) $\dfrac{12}{36} = \dfrac{1}{3}$.

3. (a) $\{0, 1, 2, 3, 4, 5\}$ and $\{1, 2, 3, 4, 5, 6\}$.

(b)

k	0	1	2	3	4	5
$P(D = k)$	$\frac{6}{36}$	$\frac{10}{36}$	$\frac{8}{36}$	$\frac{6}{36}$	$\frac{4}{36}$	$\frac{2}{36}$

k	1	2	3	4	5	6
$P(M = k)$	$\frac{1}{36}$	$\frac{3}{36}$	$\frac{5}{36}$	$\frac{7}{36}$	$\frac{9}{36}$	$\frac{11}{36}$

(c) $P(D \le 1) = \dfrac{4}{9}$, $P(M \le 3) = \dfrac{1}{4}$, $P(D \le 1 \text{ and } M \le 3) = \dfrac{7}{36}$.

(d) No. Use (c) and note that $\dfrac{7}{36} \ne \dfrac{4}{9} \cdot \dfrac{1}{4}$.

4.

$$F_D(y) = \begin{cases} 0 & y < 0 \\ \frac{6}{36} & 0 \le y < 1 \\ \frac{16}{36} & 1 \le y < 2 \\ \frac{24}{36} & 2 \le y < 3 \\ \frac{30}{36} & 3 \le y < 4 \\ \frac{34}{36} & 4 \le y < 5 \\ 1 & 5 \le y \end{cases}$$

$$F_M(y) = \begin{cases} 0 & y < 1 \\ \frac{1}{36} & 1 \le y < 2 \\ \frac{4}{36} & 2 \le y < 3 \\ \frac{9}{36} & 3 \le y < 4 \\ \frac{16}{36} & 4 \le y < 5 \\ \frac{25}{36} & 5 \le y < 6 \\ 1 & 6 \le y \end{cases}$$

5. (a) $\{1, 2, 3, 4, 5, 6, 8, 9, 10, 12, 15, 16, 18, 20, 24, 25, 30, 36\}$.

(b) $\dfrac{1}{12}$. (c) $\dfrac{1}{9}$.

6. (a) $\dfrac{1}{8}$. (b) $\dfrac{1}{4} + \dfrac{1}{2} - \dfrac{1}{8} = \dfrac{5}{8}$. (c) $\dfrac{1}{2} \cdot \dfrac{3}{4} = \dfrac{3}{8}$.

7. (a) {0, 1, 2, 3, 4}.

(b) Since $\{X + Y = 2\} = \{X = 0, Y = 2\} \cup \{X = 1, Y = 1\} \cup \{X = 2, Y = 0\}$, the union is a disjoint union and the sets are independent, we have

$$P(X + Y = 2) = \frac{1}{4} \cdot \frac{1}{2} + \frac{1}{4} \cdot \frac{1}{4} + \frac{1}{2} \cdot \frac{1}{4} = \frac{5}{16}.$$

(c)

k	0	1	2	3	4
P(X + Y = k)	$\frac{1}{16}$	$\frac{1}{8}$	$\frac{5}{16}$	$\frac{1}{4}$	$\frac{1}{4}$

8. (a)

k	2	5	8
P(3X + 2 = k)	$\frac{1}{4}$	$\frac{1}{4}$	$\frac{1}{2}$

(b)

k	0	1	2
P(2 - X = k)	$\frac{1}{2}$	$\frac{1}{4}$	$\frac{1}{4}$

9. (a) Sample calculation: $P(X = 2) = \dfrac{\binom{5}{2} \cdot \binom{5}{2}}{\binom{10}{4}} = \dfrac{10 \cdot 10}{210} = \dfrac{20}{42}.$

k	0	1	2	3	4
P(X = k)	$\frac{1}{42}$	$\frac{10}{42}$	$\frac{20}{42}$	$\frac{10}{42}$	$\frac{1}{42}$

(b) Sample calculation: $P(X = 2) = \dfrac{\binom{5}{2} \cdot \binom{5}{5}}{\binom{10}{7}} = \dfrac{10}{120} = \dfrac{1}{12}.$

k	2	3	4	5
P(X = k)	$\frac{1}{12}$	$\frac{5}{12}$	$\frac{5}{12}$	$\frac{1}{12}$

10.

k	-1	5
P(W = k)	$\frac{5}{6}$	$\frac{1}{6}$

11. (a) No. The values f(x) must sum to 1.

(b) Yes. For example, it is the probability distribution for the random variable recording a random number selected from {1, 2, 3, 4, 5}.

12.

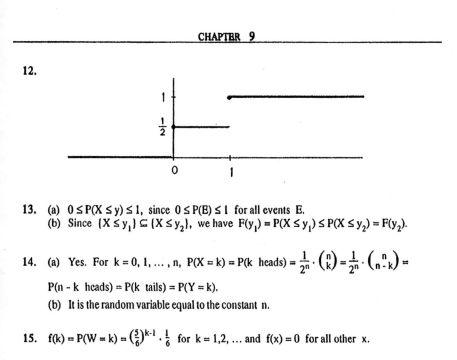

13. (a) $0 \le P(X \le y) \le 1$, since $0 \le P(E) \le 1$ for all events E.

(b) Since $\{X \le y_1\} \subseteq \{X \le y_2\}$, we have $F(y_1) = P(X \le y_1) \le P(X \le y_2) = F(y_2)$.

14. (a) Yes. For $k = 0, 1, \ldots, n$, $P(X = k) = P(k \text{ heads}) = \frac{1}{2^n} \cdot \binom{n}{k} = \frac{1}{2^n} \cdot \binom{n}{n-k} =$

$P(n - k \text{ heads}) = P(k \text{ tails}) = P(Y = k)$.

(b) It is the random variable equal to the constant n.

15. $f(k) = P(W = k) = \left(\frac{5}{6}\right)^{k-1} \cdot \frac{1}{6}$ for $k = 1,2, \ldots$ and $f(x) = 0$ for all other x.

16. (a) $\frac{1}{2^n}$. (b) $\frac{1}{3^n}$. (c) $\frac{1}{6^n}$.

17. (a) $P\left(\left[\frac{1}{6}, \frac{5}{6}\right]\right) = \frac{2}{3}$. (b) $P\left(\left[0, \frac{1}{3}\right] \cup \left[\frac{2}{3}, 1\right)\right) = \frac{2}{3}$.

18. Suppose that E and F are independent. Since the value sets of χ_E and χ_F are both

$\{0, 1\}$, we only need to check that

$$P(\chi_E = x \text{ and } \chi_F = y) = P(\chi_E = x) \cdot P(\chi_F = y) \quad \text{for } x,y \in \{0, 1\}.$$

This follows from the fact that the following pairs of events are independent: $\{E, F\}$,

$\{E, F^c\}$, $\{E^c, F\}$ and $\{E^c, F^c\}$; see Exercise 20, § 9.1. For example,

$P(\chi_E = 0 \text{ and } \chi_F = 1) = P(E^c \cap F) = P(E^c) \cdot P(F) = P(\chi_E = 0) \cdot P(\chi_F = 1)$.

Conversely, if χ_E and χ_F are independent, then

$P(E \cap F) = P(\chi_E = 1 \text{ and } \chi_F = 1) = P(\chi_E = 1) \cdot P(\chi_F = 1) = P(E) \cdot P(F)$.

19. Let $x_1, x_2, \ldots, x_m$ and $y_1, y_2, \ldots, y_n$ be the value sets for X and Y.

$(I_2) \Rightarrow (I_3)$. To show (I_3), we may assume that $x = x_i$ and $y = y_j$ for some i and j.

Let I be an interval containing x_i and no other number in X's value set, and let J be an

interval containing y_j and no other number in Y's value set. Then

$P(X = x_i \text{ and } Y = y_j) = P(X \in I \text{ and } Y \in J) = P(X \in I) \cdot P(Y \in J) = P(X = x_i) \cdot P(Y = y_j)$.

$(I_3) \Rightarrow (I_2)$. Given intervals I and J, let $A = \{i : x_i \in I\}$ and $B = \{j : y_j \in J\}$. Then $\{X \in I \text{ and } Y \in J\}$ is the disjoint union $\bigcup\limits_{i \in A} \bigcup\limits_{j \in B} \{X = x_i \text{ and } Y = y_j\}$, and so

$$P(X \in I \text{ and } Y \in J) = \sum_{i \in A} \sum_{j \in B} P(X = x_i \text{ and } Y = y_j) = \sum_{i \in A} \sum_{j \in B} P(X = x_i) \cdot P(Y = y_j)$$

$$= \left(\sum_{i \in A} P(X = x_i) \right) \cdot \left(\sum_{j \in B} P(Y = y_j) \right) = P(X \in I) \cdot P(Y \in J).$$

Section 9.3

1. (a) The answer is rather obviously 2. To confirm this, use the probability distribution in the answer to Exercise 9, § 9.2, and calculate $\sum\limits_{k=0}^{4} k \cdot P(X = k) = 2$.

   (b) $E(X^2) = \sum\limits_{k=0}^{4} k^2 \cdot P(X = k) = \frac{196}{42} = \frac{14}{3}$, so $V(X) = E(X^2) - \mu^2 = \frac{14}{3} - 4 = \frac{2}{3}$. Hence $\sigma = \sqrt{\frac{2}{3}} \approx .82$.

2. Of course there are no average families. In general, the mean of a random variable need not be in the value set of the random variable. For instance, in Example 1 the mean is 3.5, but the value on the die is never 3.5.

3. (a) mean deviation $= \sum\limits_{k=1}^{6} |k - 3.5| \cdot \frac{1}{6} = 1.5 < 1.71 \approx \sigma$.

   (b) mean deviation $= |0 - \frac{1}{2}| \cdot \frac{1}{2} + |1 - \frac{1}{2}| \cdot \frac{1}{2} = \frac{1}{2} = \sigma$.

4. $(-1) \cdot (.9999995) + 1,000,000 \cdot (.0000005) \approx -\$.50$.

5. $\mu_X = \mu_Y = \sum\limits_{k=0}^{2} k \cdot f(k) = \frac{5}{4}$ and $\mu_{X+Y} = \frac{5}{2}$. Since $E(X^2) = \sum\limits_{k=0}^{2} k^2 \cdot f(k) = \frac{9}{4}$, we have $V(X) = \frac{9}{4} - \left(\frac{5}{4}\right)^2 = \frac{11}{16}$ and $\sigma_X = \frac{1}{4}\sqrt{11} \approx .83$. Same for Y. Finally, by Theorem 7, $V(X + Y) = V(X) + V(Y) = \frac{11}{8}$ and so $\sigma_{X+Y} = \sqrt{\frac{11}{8}} \approx 1.17$.

6. Use the probability distributions from Exercise 3, § 9.2. $\mu_D = \sum\limits_{k=0}^{5} k \cdot P(D = k) = \frac{35}{18}$ and $\mu_M = \sum\limits_{k=1}^{6} k \cdot P(M = k) = \frac{161}{36} \approx 4.47$. Since $E(D^2) = \sum\limits_{k=0}^{5} k^2 \cdot P(D = k) = \frac{210}{36}$, we have $V(D) = \frac{210}{36} - \left(\frac{35}{18}\right)^2 = \frac{665}{324} \approx 2.05$ and so $\sigma_D \approx \sqrt{2.05} \approx 1.43$. Likewise $E(M^2) = \sum\limits_{k=1}^{6} k^2 \cdot P(D = k) = \frac{791}{36} \approx 21.97$, so $V(M) = \frac{791}{36} - \left(\frac{161}{36}\right)^2 \approx 1.97$ and $\sigma_M \approx 1.40$.

7. (a) $\frac{3}{5}$.   (b) $\frac{7}{5}$.   (c) $\frac{13}{5}$.   (d) $\frac{19}{5}$.

8. (a) $E(X^2) = \frac{13}{5}$, so $V(X) = \frac{13}{5} - \left(\frac{3}{5}\right)^2 = \frac{56}{25}$ and $\sigma_X = \frac{1}{5}\sqrt{56} \approx 1.50$.

    (b) $E(|X|^2) = E(X^2) = \frac{13}{5}$, so $V(|X|) = \frac{13}{5} - \left(\frac{7}{5}\right)^2 = \frac{16}{25}$ and $\sigma_{|X|} = \frac{4}{5} = .80$.

9. Since $E(X^4) = (-2)^4 \cdot \frac{1}{5} + 1^4 \cdot \frac{1}{5} + 2^4 \cdot \frac{2}{5} = \frac{49}{5}$, we have $V(X^2) = E(X^4) - [E(X^2)]^2 =$ $\frac{49}{5} - \left(\frac{13}{5}\right)^2 = \frac{76}{25}$. Hence the standard deviation of X^2 is $\frac{1}{5}\sqrt{76} \approx 1.74$.

10. The mathematical expectation is \$500,000 in the first case and \$600,000 in the second case. From the strict point of view of maximizing expectation, one should prefer the 20 percent chance of winning \$3,000,000. However, most of us would take the first choice because we regard the value of winning \$1,000,000 as much more than one third of the value of winning \$3,000,000. In other words, we don't consider the dollars after the first million to be as valuable as the first million. A theory of "utility" takes this weighting of values into account.

11. The random variable X is the sum $X_1 + X_2 + X_3 + X_4 + X_5$ where $X_i = 1$ if the ith card is an ace and $X_i = 0$ otherwise. Since $E(X_i) = \frac{1}{13}$ for each i, we have $E(X) = \frac{5}{13}$. Of course, the direct assault will also work: $E(X) = \sum_{k=1}^{4} k \cdot P(X = k)$ where

$$P(X = k) = \frac{\binom{4}{k} \cdot \binom{48}{5-k}}{\binom{52}{5}}.$$

12. You might nonsensically expect $\frac{1}{6}$ of a success at each toss, so one success in 6 tosses. Indeed, for W in Exercise 15, § 9.2, one can show that $E(W) = \sum_{k=1}^{\infty} k \cdot \left(\frac{5}{6}\right)^{k-1} \cdot \frac{1}{6} = 6$.

13. (a) Let W be the waiting time random variable. If we imagine that all five marbles are drawn from the urn, then it is clear that the blue marble is as likely to be the first marble as it is the second marble, etc. That is, $P(W = k) = \frac{1}{5}$ for $k = 1, 2, 3, 4, 5$. [These equalities can also be easily verified directly. For example, $P(W = 3) = \frac{4}{5} \cdot \frac{3}{4} \cdot \frac{1}{3} = \frac{1}{5}$.] Thus $E(W) = \frac{1}{5}(1 + 2 + 3 + 4 + 5) = 3$.

(b) The answer is 5 using the "nonsensical" argument that $\frac{1}{5}$ of a blue marble is expected on each draw. To verify this mathematically, let W be the waiting time random variable so that $P(W = k) = \left(\frac{4}{5}\right)^{k-1}\left(\frac{1}{5}\right)$ for $k \geq 1$. Then $E(W) = \sum_{k=1}^{\infty} k \cdot \left(\frac{4}{5}\right)^{k-1}\left(\frac{1}{5}\right)$, which turns out to equal 5.

14. By Theorem 6, $\frac{n}{4} = V(S_n) = E(S_n^2) - \left(\frac{n}{2}\right)^2$ or $E(S_n^2) = \frac{n + n^2}{4}$. Since $P(S_n = k) = 2^{-n}\binom{n}{k}$ for $k = 0, 1, \ldots, n$, the corollary to Theorem 3 shows that $E(S_n^2) = \sum_{k=0}^{n} k^2 \cdot 2^{-n}\binom{n}{k}$.

Hence $\sum_{k=0}^{n} k^2 \cdot \binom{n}{k} = 2^n \cdot \frac{n^2 + n}{4} = n(n + 1) \cdot 2^{n-2}$.

15. (a) Y also takes each value $1, 2, \ldots, n$ with probability $\frac{1}{n}$.

(b) $E(X) + E(Y) = E(X + Y) = E(n + 1) = n + 1$. Since $E(X) = E(Y)$ by part (a), $2 \cdot E(X) = n + 1$.

(c) By Theorem 2, $E(X) = \frac{1}{n} + \frac{2}{n} + \cdots + \frac{n}{n}$. So $1 + 2 + \cdots + n = n \cdot E(X) = n \cdot \frac{1}{2}(n + 1)$.

16. Almost any choice of non-independent random variables X and Y will do. As a simple example, toss one fair coin, let $X = 1$ if a head appears and $X = 0$ otherwise, and let $Y = 1$ if a tail appears and $Y = 0$ otherwise. Then $XY = 0$ but $E(X) \cdot E(Y) = \frac{1}{4}$.

17. If $\mu = E(X)$, then $E(X + c) = \mu + c$. Hence

$$V(X + c) = \sum_{X} (x - \mu - c)^2 \cdot P(X + c = x) = \sum_{X} (x - c - \mu)^2 \cdot P(X = x - c).$$

Replace each $x - c$ by y to get $\sum_{y} (y - \mu)^2 \cdot P(X = y) = V(X)$. The result just shown is intuitively obvious: $X + c$ shifts all values of X by c, but does not modify the spread of its values.

Since $V(cX) = c^2 \cdot V(X)$ is obvious for $c = 0$, assume $c \neq 0$. Since $E(cX) = c \cdot \mu$,

$$V(cX) = \sum_{X} (x - c \cdot \mu)^2 \cdot P(cX = x) = c^2 \sum_{X} \left(\frac{x}{c} - \mu\right)^2 \cdot P\left(X = \frac{x}{c}\right) =$$

$$c^2 \sum_{y} (y - \mu)^2 \cdot P(X = y) = c^2 \cdot V(X).$$

18. The mean of $-X$ is $-\mu$. The standard deviation of $-X$ is σ since $V(-X) = V(X)$ by Exercise 17. This should be intuitively obvious: the values of $-X$ are spread out just like those of X, but in reverse order.

19. (a) $E(S) = n \cdot \mu$. From Theorem 7, $V(S) = \sum_{i=1}^{n} V(X_i) = \sum_{i=1}^{n} \sigma^2 = n \cdot \sigma^2$; hence $\sigma_S = \sqrt{n} \cdot \sigma$.

    (b) $E(\frac{1}{n} S) = \mu$. From Exercise 17, $V(\frac{1}{n} S) = \frac{1}{n^2} V(S) = \frac{1}{n} \cdot \sigma^2$; hence the standard deviation of $\frac{1}{n} S$ is $\frac{1}{\sqrt{n}} \cdot \sigma$.

20. (a) $E(X) = \sum_{\omega \in \Omega} X(\omega) \cdot P(\{\omega\}) \le \sum_{\omega \in \Omega} Y(\omega) \cdot P(\{\omega\}) = E(Y)$.

    (b) Apply part (a) to the random variable $(X - \mu)^2$ and the constant random variable $(X(\omega_0) - \mu)^2$ to obtain $\sigma_X^2 = E((X - \mu)^2) \le E((X(\omega_0) - \mu)^2) = (X(\omega_0) - \mu)^2$. Take square roots.

21. (a) Since all these random variables have finite value sets, it suffices to show

    (1) $\quad P(X_1 + X_2 = x \text{ and } X_i = x_i \text{ for } i = 3, \ldots, n) = P(X_1 + X_2 = x) \cdot \prod_{i=3}^{n} P(X_i = x_i)$

    for real numbers $x, x_3, \ldots, x_n$. Let A be the set of all pairs (u,v) of real numbers so that u is in the value set of X_1, v is in the value set of X_2, and $u + v = x$. Then $\{X_1 + X_2 = x\}$ is the disjoint union $\bigcup_{(u,v) \in A} \{X_1 = u \text{ and } X_2 = v\}$, and so

    (2) $\quad\quad P(X_1 + X_2 = x) = \sum_{(u,v) \in A} P(X_1 = u \text{ and } X_2 = v)$.

    Similarly
    $P(X_1 + X_2 = x \text{ and } X_i = x_i \text{ for } i = 3, \ldots, n)$

    $\quad\quad = \sum_{(u,v) \in A} P(X_1 = u, X_2 = v \text{ and } X_i = x_i \text{ for } i = 3, \ldots, n)$.

    Since $X_1, X_2, \ldots, X_n$ are independent, this last sum is equal to

    $\quad\quad \sum_{(u,v) \in A} P(X_1 = u) \cdot P(X_2 = v) \cdot \prod_{i=3}^{n} P(X_i = x_i)$.

    By (2) this equals $P(X_1 + X_2 = x) \cdot \prod_{i=3}^{n} P(X_i = x_i)$ and so (1) holds.

    (b) Induction on n; the case $n = 2$ is shown in Theorem 7. Assume true for $n - 1$ independent random variables, and suppose $X_1, X_2, \ldots, X_n$ are independent. By the inductive assumption and part (a),
    $V(X_1 + X_2 + \cdots + X_n) = V(X_1 + X_2) + V(X_3) + \cdots + V(X_n)$.
    Finally $V(X_1 + X_2) = V(X_1) + V(X_2)$ by the $n = 2$ case.

Section 9.4

1. If the outcomes were not independent, we would have to use conditional probabilities to conclude that the probability of (S,S,F,S,F) is

   P(first is S)·P(second is S | first is S)·P(third is F | first and second are S)·etc.

2. (a) Expected number is $np = 1$, since $n = 3$ and $p = \frac{1}{3}$.

   (b) $1 - P(\text{no hits}) = 1 - \left(\frac{2}{3}\right)^3 \approx .704$.

3. (a) Expected number is $np = \frac{10}{3}$, since $n = 10$ and $p = \frac{1}{3}$.

   (b) $F(3) \approx .559$ from Table 1.

   (c) $1 - P(\text{at most } 2 \text{ hits}) = 1 - F(2) \approx 1 - .299 = .701$, using Table 1.

4. (a) $\frac{1}{2^{10}} \cdot \binom{10}{5}$. This is $F(5) - F(4) \approx .623 - .377 = .246$ from Table 1.

   (b) $1 - F(2) \approx 1 - .055 = .945$.

5. (a) $(.9)^{10} \approx .349$ or $F(0) \approx .349$ from Table 1.

   (b) $F(2) \approx .930$ from Table 1.

6. (a) $\Phi(1) - \Phi(-1) \approx .68$.   (b) $\Phi(2) - \Phi(-2) \approx .95$.

   (c) $\Phi(3) - \Phi(-3) \approx .997$.

7. (a) $1 - \Phi(1) \approx 1 - .8413 = .1587$.   (b) $\Phi(-2) \approx .0227$.

8. $\sigma = \sqrt{30000 \cdot \frac{1}{4} \cdot \frac{3}{4}} = 75$, and so $P(7400 < X \leq 7600)$

   $\approx P(\mu - 1.33\sigma < X \leq \mu + 1.33\sigma) \approx \Phi(1.33) - \Phi(-1.33)$. This turns out to be approximately .818.

9. (a) $\mu = 600$.   (b) $\sigma^2 = 1800 \cdot \frac{1}{3} \cdot \frac{2}{3} = 400$, so $\sigma = 20$.

   (c) As in Example 9, one such interval is $(\mu - 2\sigma, \mu + 2\sigma] = (560, 640]$.

10. Since $\Phi(2) \approx .977$, he should plan on about $\mu + 2\sigma$ folks. Here $\mu = 600$ and $\sigma^2 = 1000(.60)(.40) = 240$. So he should plan on $600 + 2 \cdot \sqrt{240} \approx 631$ people.

11. (a) $\mu = 500$ and $\sigma = \sqrt{1000 \cdot \frac{1}{2} \cdot \frac{1}{2}} \approx 15.81$. Thus $10 \approx .632 \cdot \sigma$ and so

    $P(490 < X \leq 510) \approx P(\mu - .63\sigma < X \leq \mu + .63\sigma) \approx \Phi(.63) - \Phi(-.63)$.

This turns out to be approximately .47.

(b) $\mu = 5000$ and $\sigma = 50$, so $P(4900 < X \le 5100)$
$= P(\mu - 2\sigma < X \le \mu + 2\sigma) \approx \Phi(2) - \Phi(-2) \approx .95$.

(c) $\mu = 500,000$ and $\sigma = 500$, so $P(490,000 < X \le 510,000)$
$= P(\mu - 20\sigma < X \le \mu + 20\sigma) \approx \Phi(20) - \Phi(-20)$.

This is very, very close to 1: .9999⋯ where the first 88 digits are 9. For all practical purposes, the event $\{490,000 < X \le 510,000\}$ is a certainty.

12. Some of the students must have understood the question. If the question had been answered at random by all students, the number X of correct answers would be a binomial random variable with $n = 1,000,000$ and $p = \frac{1}{4}$. Here $\mu = 500,000$ and $\sigma = 500$, just as in Exercise 11. So $P(X \ge 510,000) \approx 1 - \Phi(20)$. This is .0000⋯ with about 88 zeros [compare Exercise 11(c)], so the event $\{X \ge 510,000\}$ is *extremely* unlikely. It is possible that about 2% of the students really understood the question and that the other 98% guessed at random, but the truth is probably more complicated than this.

13. (a) $\frac{13}{12}$. Note that $X = X_1 + X_2 + X_3$ where $X_i = 1$ if the ith experiment is a success and $X_i = 0$ otherwise. So $E(X) = E(X_1) + E(X_2) + E(X_3) = \frac{1}{2} + \frac{1}{3} + \frac{1}{4}$.

(b) Since the experiments are independent, Theorem 7 of § 9.3 implies that
$$V(X) = V(X_1) + V(X_2) + V(X_3) = \frac{1}{2} \cdot \frac{1}{2} + \frac{1}{3} \cdot \frac{2}{3} + \frac{1}{4} \cdot \frac{3}{4} = \frac{95}{144}.$$
Hence $\sigma_X = \frac{1}{12} \sqrt{95} \approx .81$.

(c) No; the probabilities of successive experiments are not the same fixed value p.

14. By Theorem 2(b), $F_2(y) = \tilde{F}_2(\frac{y - \mu_2}{\sigma_2}) = \tilde{F}_1(\frac{y - \mu_2}{\sigma_2})$. Now apply Theorem 2(c) to obtain
$$F_2(y) = F_1(\sigma_1 \cdot (\frac{y - \mu_2}{\sigma_2}) + \mu_1).$$

15. (a) Since y or -y is nonnegative, we may assume that one of them, say y, is nonnegative. Now

    1 = area under the bell curve φ

       = area under φ to left of y + area under φ to right of y

       = $\Phi(y)$ + area under φ to the right of y.

By symmetry of the graph of φ,

the area under φ to the right of y = the area under φ to the left of $-y$ = $\Phi(-y)$, and so $1 = \Phi(y) + \Phi(-y)$.

(b) By part (a), $\Phi(-y) = 1 - \Phi(y)$, so $\Phi(y) - \Phi(-y) = \Phi(y) - [1 - \Phi(y)] = 2 \cdot \Phi(y) - 1$.

16. (a) By Theorem 2(b), $P(\mu - \sigma < X \leq \mu + \sigma) = F(\mu + \sigma) - F(\mu - \sigma) =$
$\tilde{F}(\frac{\mu + \sigma - \mu}{\sigma}) - \tilde{F}(\frac{\mu - \sigma - \mu}{\sigma}) = \tilde{F}(1) - \tilde{F}(-1) = \Phi(1) - \Phi(-1)$.

(b) $P(\mu - c\sigma < X \leq \mu + c\sigma) = F(\mu + c\sigma) - F(\mu - c\sigma) = \tilde{F}(\frac{\mu + c\sigma - \mu}{\sigma}) - \tilde{F}(\frac{\mu - c\sigma - \mu}{\sigma}) =$
$\tilde{F}(c) - \tilde{F}(-c) = \Phi(c) - \Phi(-c)$. This can also be written as $2 \cdot \Phi(c) - 1$ by Exercise 15.

CHAPTER 10

We view Boolean algebras as algebraic structures with operations ∧, ∨ and '. $\mathscr{P}(S)$ is the familiar model and FUN(S,$\mathbb{B}$) is the example we want to understand better. The natural order relation ≤ is derived from the operations in the same way that ⊆ can be expressed in terms of ∩ or ∪. The key idea is the unique expression as joins of atoms, which we then exploit to talk about Boolean functions [i.e., switching functions]. After a brief digression to look at logical circuits [= networks] we describe the Karnaugh map procedures for finding Boolean expressions which are in some sense "optimal."

Section 10.1 contains the algebraic facts. At first it's not clear where all of this is leading. Tell the students that our aim is to describe logical circuits algebraically. Theorem 3 shows why we care so much about atoms; they are the basic building blocks for [finite] Boolean algebras. Theorem 4 says that if you've seen one Boolean algebra with n atoms you've seen 'em all. The notion of Boolean function as illustrated in Example 6 will be important in what follows. The link with truth tables provides the connection with logic. Wright, the algebraist, thinks Hasse diagrams are extremely valuable. Exercise 9 gives a chance to discuss them if you don't plan to cover § 11.1. Exercise 14 is fairly tough for the students to do, but makes a reasonable classroom example. Ross, the analyst, likes it.

The key idea in § 10.2 is that Boolean expressions that look different may produce the same Boolean function. We want to choose a "good" representative expression for each Boolean function. Work out an example similar to Example 6 to illustrate terminology and the connection between the values in the literal columns and the values in the final column. Perhaps carry out the details of Example 7(b). See also the answer and comments below on Exercise 9.

Section 10.3 relates Boolean expressions to hardware. The challenge thus becomes: find the "best" hardware configuration to produce a given logical [= Boolean] function. We don't pretend to give a complete answer, since to do so would be difficult even if we knew all of the associated costs and constraints, but we hope to give the flavor of the problem. Students seem to find this material concrete and enjoyable. Many will have seen it in their computer science courses. The idea in Example 5 can be amplified to discuss time complexity versus space complexity in computation.

The idea behind the Karnaugh maps in § 10.4 seems easy for students to grasp. The difficulty lies in making sure that we've chosen the best set of blocks. The erroneous "solution" in

Figure 4(b) was obtained using the method in a competitor's book. This example illustrates the fact that the choice process must be somewhat sophisticated.

Section 10.1

1. (a) Since the operations $\vee$ and $\wedge$ treat 0 and 1 just as if they represent truth values, checking the laws 1Ba through 5Bb for all cases amounts to checking corresponding truth tables. Do enough until the situation is clear to you.

    (b) We illustrate laws 3Ba and 5Ba. For each $x \in S$
    $$(f \vee (g \wedge h))(x) = f(x) \vee (g \wedge h)(x) = f(x) \vee [g(x) \wedge h(x)]$$
    by definition of $\vee$ and $\wedge$ on FUN(S,$\mathbb{B}$). Likewise,
    $$((f \vee g) \wedge (f \vee h))(x) = [f(x) \vee g(x)] \wedge [f(x) \vee h(x)].$$
    Since $\mathbb{B}$ is a Boolean algebra, we have
    $$f(x) \vee [g(x) \wedge h(x)] = [f(x) \vee g(x)] \wedge [f(x) \vee h(x)] \quad \text{for all } x \in S.$$
    So $f \vee (g \wedge h) = (f \vee g) \wedge (f \vee h)$. For each $x \in S$, $(f \vee f')(x) = f(x) \vee f'(x) = f(x) \vee f(x)' = 1$ and so $f \vee f' = 1$, the function identically 1 on S.

2. (a) $\{a, c, d\} = \{a\} \cup \{c\} \cup \{d\}$.

    (b) $(1,0,1,1,0) = (1,0,0,0,0) \vee (0,0,1,0,0) \vee (0,0,0,1,0)$.

    (c) $f = f_a \vee f_c \vee f_d$ where f_a, f_c, f_d are defined as follows

	a	b	c	d	e
f_a	1	0	0	0	0
f_c	0	0	1	0	0
f_d	0	0	0	1	0

    Note the similarity among parts (a), (b) and (c).

3. One solution is to set $S = \{1, 2, 3, 4, 5\}$ and define $\varphi(x_1,x_2,x_3,x_4,x_5) = \{i \in S : x_i = 1\}$.

4. They are the functions that take the value 1 at exactly one element in S. This is true whether S is finite or not.

5. (a) The atoms are given by the four columns on the right in the table.

x	y	a	b	c	d
0	0	1	0	0	0
0	1	0	1	0	0
1	0	0	0	1	0
1	1	0	0	0	1

(b) In the notation of the answer to (a), $g = c \vee d$.

(c) In the notation of the answer to (a), $h = a \vee b \vee d$.

6. (a) $2^4 = 16$. (b) 5. (c) $\binom{16}{5} = 4368$.

7. (a) No. A finite Boolean algebra has 2^n elements for some n.

(b) No, since $|BOOL(n)| = 2^{2^n}$. For example, 8 does not have this form.

8. (a) The sets $\{n\}$, $n \in \mathbb{N}$.

(b) No, since we only take the join of a finite collection of elements. In our case, the join (union!) of atoms will always be a finite subset of $\mathbb{N}$.

9. (a)   (b)

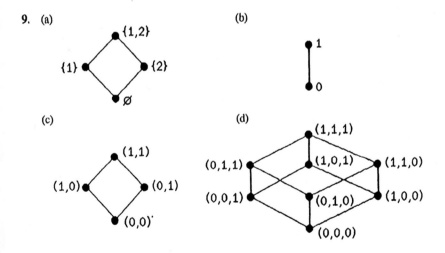

(c)   (d)

10. (a) The 0 element is the integer 1; 6 is the 1 element.

(b) D_6 and $\mathscr{P}(\{2, 3\})$ are isomorphic, using φ defined by $\varphi(1) = \emptyset$, $\varphi(2) = \{2\}$,

$\varphi(3) = \{3\}$ and $\varphi(6) = \{2, 3\}$. Any other 2-element set would work as well as $\{2, 3\}$, but the association with $\{2, 3\}$ is especially obvious.

(c) $D_4 = \{1, 2, 4\}$ has 3 elements, while finite Boolean algebras must have 2^n elements for some $n \in \mathbb{P}$.

(d) D_8 has 4 elements. However, 2 and 4 have no complements in D_8. For example, if $2' = z$ then $2 \wedge z = 1$ implies $z = 1$ while $2 \vee z = 8$ implies $z = 8$.

11. (a) If $a \le x$ or $a \le y$, then surely $a \le x \vee y$ by Lemmas 3(a) and 2(a). Suppose $a \le x \vee y$. Then $a = a \wedge (x \vee y) = (a \wedge x) \vee (a \wedge y)$. One of $a \wedge x$ and $a \wedge y$, say $a \wedge x$, must be different from 0. But $0 < a \wedge x \le a$, so $a \wedge x = a$ and $a \le x$.

(b) If $a \le x$ and $a \le y$, then $a = a \wedge y = (a \wedge x) \wedge y = a \wedge (x \wedge y)$, so $a \le (x \wedge y)$. If $a \le (x \wedge y)$, then clearly $a \le x$ and $a \le y$.

(c) $a \le 1 = x \vee x'$, so $a \le x$ or $a \le x'$ by part (a). Both $a \le x$ and $a \le x'$ would imply $a \le x \wedge x' = 0$ by part (b), a contradiction.

12. (a) $x \vee y = a_1 \vee \cdots \vee a_n \vee b_1 \vee \cdots \vee b_m$; remove duplicates. $x \wedge y$ is the join of the atoms used for both x and y. For example, in $\mathcal{P}(\{1, 2, 3\})$ we have $\{1, 2\} = \{1\} \cup \{2\}$ and $\{2, 3\} = \{2\} \cup \{3\}$, so $\{1, 2\} \cup \{2, 3\} = \{1\} \cup \{2\} \cup \{3\}$ and $\{1, 2\} \cap \{2, 3\} = \{2\}$.

(b) Use exactly the set of atoms not used by x. For example, in $\mathbb{B}^4$, $(1,1,0,0) = (1,0,0,0) \vee (0,1,0,0)$ so $(1,1,0,0)' = (0,0,1,0) \vee (0,0,0,1) = (0,0,1,1)$.

13. $x \le y \Leftrightarrow x \vee y = y \Leftrightarrow \varphi(x \vee y) = \varphi(y) \Leftrightarrow \varphi(x) \vee \varphi(y) = \varphi(y) \Leftrightarrow \varphi(x) \le \varphi(y)$.

14. (a) It suffices to consider $\{[a_i, b_i)\}_{i \in \mathbb{P}}$ and show that each union $\bigcup_{i=1}^{n} [a_i, b_i)$ can be written as a finite disjoint union of such intervals. This is obvious for $n = 1$. Assume it has been done for $\bigcup_{i=1}^{n-1} [a_i, b_i)$; in fact, we can assume the intervals $[a_1, b_1), \ldots, [a_{n-1}, b_{n-1})$ are disjoint. If $[a_n, b_n)$ intersects none of these sets, we're done. Otherwise, let

$I = \{i : 1 \le i \le n-1 \text{ and } [a_i, b_i) \cap [a_n, b_n) \ne \emptyset\}$,

$a^* = \min(\{a_i : i \in I\} \cup \{a_n\})$,

$b^* = \max(\{b_i : i \in I\} \cup \{b_n\})$,

and observe $[a_n, b_n) \cup \bigcup_{i \in I} [a_i, b_i) = [a^*, b^*)$ so that

$$[a^*, b^*) \cup \{[a_i, b_i) : i \notin I, \ 1 \le i \le n-1\}$$

expresses $\bigcup_{i=1}^{n} [a_i, b_i)$ as a disjoint union of intervals of the form $[a, b)$.

(b) Since $\mathcal{A}$ inherits the laws B1, ..., B5 from the Boolean algebra $\mathcal{P}(S)$, we just need to show that if X and Y are in $\mathcal{A}$ then so are $X \cup Y$, $X \cap Y$ and X'. This is clear for $\cup$ and easy for $\cap$ since

$$[a, b) \cap [c, d) = [\max\{a, c\}, \min\{b, d\})$$

whenever the intersection is nonempty. For complementation, consider a disjoint union $\bigcup_{i=1}^{n} [a_i, b_i)$ and assume that $a_1 < b_1 < a_2 < b_2 < \cdots < a_n < b_n$. Then the complement is

$$[0, a_1) \cup \left(\bigcup_{i=1}^{n-1} [b_i, a_{i+1}) \right) \cup [b_n, 1).$$

The union may be $[0, a_1)$ or $[b_n, 1)$.

(c) $\mathcal{A}$ has no atoms since nonempty members of $\mathcal{A}$ always contain smaller nonempty members of $\mathcal{A}$. For example, $[a, b)$ contains $[a, \frac{1}{2}(a + b))$.

Section 10.2

1. $x'y'z' \vee x'y'z \vee xyz'$.

2. From E's minterm canonical form we see that the corresponding $f: B^3 \to \mathbb{B}$ maps $(0,0,1)$ and $(1,1,0)$ to 0 and all other triples to 1.

3.

x	y	z	(a) xy	(b) z'	(c) xy ∨ z'	(d) 1
0	0	0	0	1	1	1
0	0	1	0	0	0	1
0	1	0	0	1	1	1
0	1	1	0	0	0	1
1	0	0	0	1	1	1
1	0	1	0	0	0	1
1	1	0	1	1	1	1
1	1	1	1	0	1	1

(a) $xyz' \vee xyz$.

(b) $x'y'z' \vee x'yz' \vee xy'z' \vee xyz'$.

(c) $x'y'z' \vee x'yz' \vee xy'z' \vee xyz' \vee xyz$.

(d) Use all eight minterms.

4. (a)

x	y	z	$x \lor yz$
0	0	0	0
0	0	1	0
0	1	0	0
0	1	1	1
1	0	0	1
1	0	1	1
1	1	0	1
1	1	1	1

(b) $x'yz \lor xy'z' \lor xy'z \lor xyz' \lor xyz$.

5. (a) $x_1 x_2 x_3' x_4 \lor x_1 x_2 x_3' x_4' \lor x_1' x_2 x_3 x_4'$.

(b) $(x_1 \lor x_2)x_3' x_4 = x_1 x_3' x_4 \lor x_2 x_3' x_4 = x_1 x_2 x_3' x_4 \lor x_1 x_2' x_3' x_4 \lor x_1 x_2 x_3' x_4 \lor x_1' x_2 x_3' x_4$

$= x_1 x_2 x_3' x_4 \lor x_1 x_2' x_3' x_4 \lor x_1' x_2 x_3' x_4$.

6. $((x \lor y)' \lor z)'$

$= (x \lor y)'' z'$ DeMorgan

$= (x \lor y)z'$ since $x'' = x$ in general

$= xz' \lor yz'$ distributive law

$= xyz' \lor xy'z' \lor xyz' \lor x'yz'$ $y \lor y' = 1, \; x \lor x' = 1$

$= xyz' \lor xy'z' \lor x'yz'$.

7. (a) $xz \lor y'$. Note $y' \lor y'z = y'z' \lor y'zv \lor y'z = y'z' \lor y'z = y'$ and similarly $y' \lor xy'z' = y'$.

(b) $xy \lor xyz = xy$ and $xz \lor z = z$, so the expression is equivalent to $xy \lor z$.

8. (a) $xyz \lor xy'z' \lor x'yz' \lor x'y'z$ (b) Same.

9. $x'y' \lor z \lor xyz \lor xy'z' = x'y'z \lor x'y'z' \lor z \lor xyz \lor xy'z'$. Now $x'y'z \lor z \lor xyz = z$ and $x'y'z' \lor xy'z' = y'z'$, so we get $z \lor y'z' = yz \lor y'z \lor y'z' = (yz \lor y'z) \lor (y'z \lor y'z') = z \lor y'$. This expression can also be obtained from a table of the corresponding Boolean function. The expression is not a single product of literals, by Exercise 12 [its function has the value 1 in more than 4 places], so any equivalent expression as a join of products of literals has at least two products, with at least one literal in each product. Thus the expression $z \lor y'$ is optimal. The purpose of Exercises 9 and 13 is to give students exercises in which they can find an optimal expression by brute force, recognize that it's optimal because the form is so simple, realize that they'd just gotten lucky and wonder if there's a general method. The exercises are intended to set them up to beg for Karnaugh maps.

162

10. $(xyz \vee xyz') \vee (xy'z \vee xyz) = xy \vee xz.$

11. (a) Find the minterm canonical form for E'. Then find $E = (E')'$ using DeMorgan laws, first on joins and then on products.

    (b) If $E = xy' \vee x'y$, then $E' = xy \vee x'y'$ in minterm canonical form. Thus
    $$E = (xy \vee x'y')' = (xy)'(x'y')' = (x' \vee y')(x \vee y).$$

12. We follow the hint: $E = E \cdot (w_1 \vee w_1') \cdots (w_{n-k} \vee w_{n-k}')$. Apply the distributive law $n - k$ times to get an expression for E as a join of 2^{n-k} minterms.

13. (a) The Boolean function for $x'z \vee y'z$ takes the value 1 at three elements in $\mathbb{B}^3$. The Boolean functions for products of literals take the value 1 at 1, 2 or 4 elements in $\mathbb{B}^3$ by Exercise 12.

    (b) The Boolean function for a single literal takes the value 1 at four elements in $\mathbb{B}^3$. Joining it with other products of literals will only increase the number of elements with value 1.

14. If f_1, f_2 are the Boolean functions for E_1, E_2, then $f_1 \vee f_2$ and $f_2 \vee f_1$ are the Boolean functions for $E_1 \vee E_2$ and $E_2 \vee E_1$, respectively. Boolean functions form a Boolean algebra, so $f_1 \vee f_2 = f_2 \vee f_1$. Hence $E_1 \vee E_2$ and $E_2 \vee E_1$ are equivalent.

Section 10.3

1. (a) $\{[((xy)z)' \vee x'] \vee [(z \vee y')']\}'$ simplifies to xyz with a little work.

   (b)
   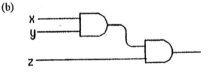

2. NOT $\quad x' = (x \vee x)'$
   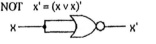

   OR $\quad x \vee y = ((x \vee y)')'$
   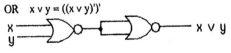

AND $x \wedge y = (x' \vee y')'$

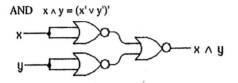

$x \wedge y$

3. (a)

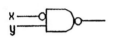

(b)

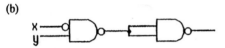

4. (a) $(x \vee y)(z \vee w) = [(x \vee y)' \vee (z \vee w)']'$.

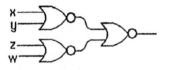

(b) $(xy')' = x' \vee y$.

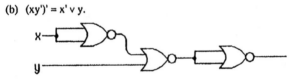

5. (a)

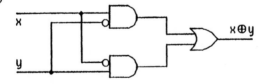

$x \oplus y$

(b) $x \oplus y = (x' \vee y')(x \vee y)$.

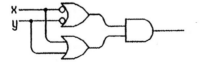

164

6. (a) $(x \oplus y \oplus z)(xyz)'$ corresponds to the network

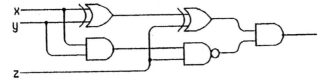

There are other possible answers, corresponding to different Boolean expressions, e.g.,

$(x \oplus y)z' \vee (x \vee y)'z$.

(b) $(x \vee y \vee z)w \vee (x \vee y)z \vee xy$ corresponds to the network

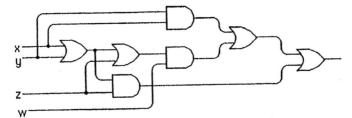

One alternative uses the Boolean expression $((xz) \vee (y \vee w))((yw) \vee (x \vee z))$.

7. (a) $S = 1$, $C_O = 0$.

(b) $S = x \oplus y \oplus C_I = 1 \oplus 1 \oplus 0 = 0$, $C_O = 1\,1 \vee (0)\,0 = 1$.

(c) $S = 0$, $C_O = 1$.

(d) $S = 1 \oplus 1 \oplus 1 = 1$, $C_O = 1$.

8. (a) $x = y = C_I = 0$.

(b) Exactly two of x, y, C_I equal 1.

(c) $x = y = C_I = 1$.

9. (a)

(b) $xy = x(y')'$.

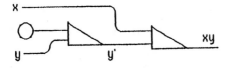

(c)

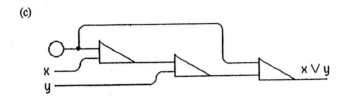

10. Since $x \oplus y = (xy \vee x'y')'$, input x' for z and y' for w.

11.

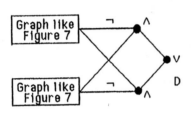

12. (a)

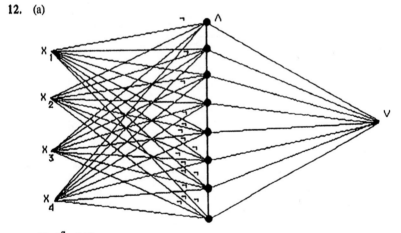

(b) $2^7 = 128$.

13. It is convenient to view the result as valid for $n = 1$. Apply the Second Principle of Mathematical Induction. A glance at a piece of Figure 7 shows that the result is valid for $n = 2$. Assume the result is true for all j with $1 \le j < n$. Consider k with $2^{k-1} < n \le 2^k$, and let $n' = 2^{k-1}$, $n'' = n - 2^{k-1}$. By the inductive assumption, there are digraphs D' and D'' for computing $x_1 \oplus x_2 \oplus \cdots \oplus x_{n'}$ and $x_{n'+1} \oplus \cdots \oplus x_n$. D' has $3(n'-1)$ $\wedge$ and $\vee$

166

vertices and D" has $3(n" - 1)$ $\wedge$ and $\vee$ vertices. Also, every path in D' and in D" has length at most $2(k - 1)$. Now create D as shown.

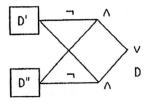

D has $3(n' - 1) + 3(n" - 1) + 3$ $\wedge$ and $\vee$ vertices, i.e., $3(n - 1)$ such vertices. Moreover, every path in D has length at most $2(k - 1) + 2 = 2k$.

14.

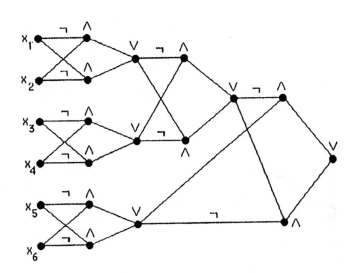

Section 10.4

1. $xyz \vee xyz' \vee xy'z' \vee xy'z \vee x'yz \vee x'y'z = x \vee z.$

2. $xyz \vee xyz' \vee xy'z \vee x'yz \vee x'y'z' \vee x'y'z = xy \vee x'y' \vee z.$

3. $xyz \vee xyz' \vee xy'z \vee x'y'z' \vee x'y'z = xz \vee xy \vee x'y' = xy \vee y'z \vee x'y'.$

4. $xyz \vee xy'z \vee x'y'z' = xz \vee x'y'z'.$

167

5. (a)

(b)

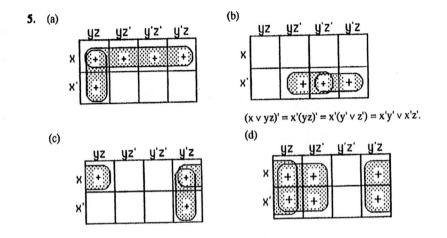

$(x \lor yz)' = x'(yz)' = x'(y' \lor z') = x'y' \lor x'z'.$

(c)

(d)

6. (a) E is a product of literals and F is a product of literals that includes the ones in E.

(b) Consider E with Boolean expression z' and F with Boolean expression xz', say.

7. (a)

(b)

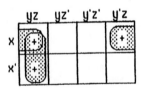

Each two-square block is essential.

Each two-square block is essential.

(c)

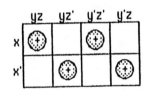

Each one-square block is essential.

8. (a)

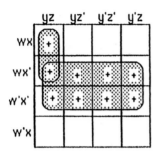

Both marked blocks are essential.

(b)

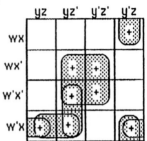

Blocks for x'z', w'xy and xy'z are essential and they cover.

(c)

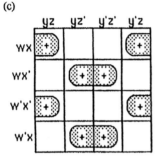

Each two-square block is essential.

9. (a) z' ∨ xy ∨ x'y' ∨ w'y or z' ∨ xy ∨ x'y' ∨ w'x'.
   (b) w'x' ∨ w'z' ∨ w'y' ∨ wxyz.
   (c) w'x'z' ∨ w'xy ∨ wxy ∨ wx'z ∨ y'z', not w'x'z' ∨ w'xy ∨ wx'y' ∨ wyz ∨ wxz', which also has five product terms but one more literal.
   (d) wz ∨ xz ∨ w'x'z'.

10. (a) wx ∨ wy ∨ wz ∨ xy ∨ xz ∨ yz. Each four-square block of the Karnaugh map is essential.
    (b) w(x ∨ y ∨ z) ∨ x(y ∨ z) ∨ yz is one example.

CHAPTER 11

The first two sections of this chapter deal with partial orders. The last two look at properties of relations in general. One can skip over the partial orders to get to the material on matrices and relations. If time permits, though, we recommend covering order relations.

Section 11.1 introduces partial orders and the basic terminology of posets. Hasse diagrams are especially important as pictorial representations. It is a good idea to present a number of examples of posets by giving their Hasse diagrams. The theorem shows the link with digraphs. There is some unavoidable overlap with Chapter 10 in order to make the chapters independent. The material on lattices can be touched on lightly. Otherwise the section will probably need two days.

The main message of § 11.2 is that chains are often nice structures, and that if $S, T, ..., U$ are chains it's possible to make $S \times T \times \cdots \times U$ into a chain in a natural way. But **not** with the product order. The right order on $S \times T \times \cdots \times U$ is the filing order. It gives an order on $\Sigma^k = \Sigma \times \cdots \times \Sigma$ as well. There are two natural orders on Σ^*, the standard order and the lexicographic order, both of which agree with the filing order on each Σ^k. They differ in the way they relate words of different lengths. Give examples to illustrate both of these orders, and assign Exercise 15. Point out that lexicographic order can have infinitely many words between two given words, while standard order has only finitely many words before any given word. If this situation distresses your students, you can illustrate the same phenomenon with the set Q^+ of positive rationals. Think of the usual order on the terminating decimals in the unit interval. Sections 11.1 and 11.2 have lots of interesting exercises. An extra day here is a good idea, if time permits.

Section 11.3 picks up from Chapter 3. The main themes are composition of relations, which is a surprisingly difficult concept for many students, and the connection between relations and Boolean matrices. Theorem 1 presents composition in matrix form. Students like to see how easily some properties can be read off at once from the matrices. One way to illustrate the trouble in spotting transitivity from a matrix is to take an example that is known to be transitive, such as a partial order, and to order its rows and columns awkwardly. Linking symmetric relations with graphs gives students a geometric handle to grasp.

The notions of transitive, reflexive and symmetric closures in § 11.4 may seem too intuitively obvious to be worth a whole section. The trick here is to raise doubts first. Look at Theorem 1, which gives the obviously correct answers, and then look at Theorem 3. Why should tsr(R) be symmetric? After all, Example 4 shows that st(R) need not be transitive. Something subtle is going on, which takes proof. Then go back to the beginning of the section for details. Exercise 9 is more trouble than fun for some students.

Section 11.1

1. (a)   (b)

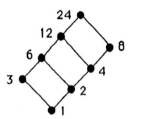 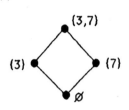

3. (a) h, o, p, q, r, z.
   (b) Posets A, B and C all have minimal elements.
   (c) B and C.   (d) f and g.   (e) f, z, p, does not exist.
   (f) Only C is a lattice. A lacks glb's and B lacks lub's.

4. They are the 2-element subsets, namely $\{a, b\}$, $\{a, c\}$ and $\{b, c\}$.

5. (a) $a \vee b = \text{lub}(a,b) = \max\{a, b\}$, $a \wedge b = \text{glb}(a,b) = \min\{a, b\}$.
   (b) $\mathbb{R}$ itself is an example.   (c) 73.   (d) 73.
   (e) $\sqrt{73}$.   (f) $-\sqrt{73}$.

6. There must be no chains A, B, C, ..., D, A in which A calls B, B calls C, ... , D calls A. A chain of this sort yields $A < D < \cdots < C < B < A$. So $A < A$ by (T), which violates (AR).

7. (a) Suppose that $\preceq$ is a partial order on S and that $\succeq$ is defined by $x \succeq y$ if and only if $y \preceq x$. Then $x \preceq x$, so $x \succeq x$. If $x \succeq y$ and $y \succeq x$, then $y \preceq x$ and $x \preceq y$, so $x = y$. If $x \succeq y$ and $y \succeq z$, then $y \preceq x$ and $z \preceq y$, so $z \preceq x$ and thus $x \succeq z$. Thus $\succeq$ satisfies (R), (AS) and (T).
   (b) Clearly $x \preceq x$, so (R) holds for $\preceq$. If $x \preceq y$ and $y \preceq z$ there are four possible cases:
   $$x = y = z, \quad x = y < z, \quad x < y = z \quad \text{and} \quad x < y < z.$$
   In the first case $x = z$. In the other three cases $x < z$. In any case $x \preceq z$. Thus (T) holds for $\preceq$.
   If $x \preceq y$ and $y \preceq x$ the cases are:

171

$x = y = x$, $\quad x = y < x$, $\quad x < y = x$ $\quad$ and $\quad x < y < x$.

Only the first is possible. The other three cases violate (AR) for $<$. Thus $\preceq$ satisfies (AS).

8. Transitivity is a general property of reachability: string together a path from u to v and a path from v to w to get a path from u to w. Antisymmetry follows from acyclicity.

9. Since $w = w\lambda$, $w \preceq w$ and (R) holds. If $w_1 \preceq w_2$ and $w_2 \preceq w_3$, there are words u and v with $w_2 = w_1 u$ and $w_3 = w_2 v$. Then $w_3 = w_1 uv$ with $uv \in \Sigma^*$ and so $w_1 \preceq w_3$. Thus (T) holds.

10. Yes. See the answer for Exercise 9 for a model argument for (R) and (T). (AS) is easily checked also: If $w_1 \preceq w_2 \preceq w_1$, then $w_2 = ww_1 w'$ and $w_1 = uw_2 u'$ for w, w', u, u' in Σ^*. Then $w_1 = uww_1 w'u'$ and this forces $u = w = w' = u' = \lambda$. So $w_1 = w_2$.

11. Not if Σ has more than one element. Show that antisymmetry fails.

12. (a) Use $x \vee y = y \vee x$ and $x \vee x = x$.

$\vee$	a	b	c	d	e	f
a	a	e	a	e	e	a
b	e	b	d	d	e	b
c	a	d	c	d	e	c
d	e	d	d	d	e	d
e	e	e	e	e	e	e
f	a	b	c	d	e	f

(b) e is the largest element and f is the smallest.

(c) Since $f \vee c = c$, $f \preceq c$. Since $c \vee d = d$, $c \preceq d$. Since $d \vee e = e$, $d \preceq e$.

(d)

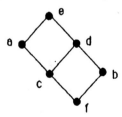

13. (a) No. Every finite subset of $\mathbb{N}$ is a subset of a larger finite subset of $\mathbb{N}$.

(b) $\varnothing$ is the unique minimal element.

(c) $\text{lub}\{A, B\} = A \cup B$. Note that $A \cup B \in \mathscr{F}(\mathbb{N})$ for all $A, B \in \mathscr{F}(\mathbb{N})$.

(d) $\text{glb}\{A, B\} = A \cap B$.　　(e) Yes; see parts (c) and (d).

14. (a) $\mathbb{N}$ is the unique maximal element.

(b) $\mathscr{I}(\mathbb{N})$ has no minimal element. Each member S of $\mathscr{I}(\mathbb{N})$ contains members of $\mathscr{I}(\mathbb{N})$ obtained by deleting finite sets from S.

(c) $\text{lub}\{A, B\} = A \cup B$.

(d) $\text{glb}\{A, B\}$ need not exist, since $A \cap B$ may be finite for infinite A and B. If $A \cap B$ *is infinite,* then $\text{glb}\{A, B\}$ exists and equals $A \cap B$.

(e) No, by part (d).

15. (a) Only $\le$. $<$ is not reflexive and $\preceq$ is not antisymmetric.

(b) Only $<$. Neither $\le$ nor $\preceq$ is antireflexive.

(c) 　　　　　　　　(d)

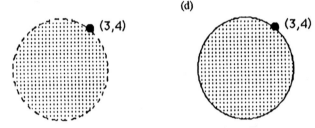

16. (a) Any subset of $\mathbb{N}$ that contains $\{1, 2, 3\}$ and has an even number of elements is an upper bound.

(b) No. For example, both $\{1, 2, 3, 4\}$ and $\{1, 2, 3, 5\}$ cover A and B. A lub of $\{A, B\}$ would have to be contained in $\{1, 2, 3, 4\} \cap \{1, 2, 3, 5\} = \{1, 2, 3\}$ and have an even number of elements.

(c) No, by part (b).

17. See Figures 2 and 7 or Exercise 16 for two different sorts of failure.

18. (a) Every poset with 1 member clearly has a minimal element. Assume inductively that every poset with $n \ge 1$ members has a minimal element, and consider a poset $(S, \preceq)$ with $n + 1$ members. Choose $x \in S$. The subposet $(S \setminus \{x\}, \preceq)$ has a minimal member, say m. Suppose first that $x \not\preceq m$. If $s \preceq x$ for some $s \in S \setminus \{x\}$, then $s \preceq m$ by (T). Then

173

$s = m$ by minimality of m, so $x \preceq m$, $m \preceq x$ and $x = m$ by (AS), a contradiction. Thus if $x \preceq m$, then x is minimal in S. But if $x \preceq m$ fails, then since m is already minimal in $S \setminus \{x\}$, m is minimal in S. The claim follows by induction.

(b) Use $(0,1]$ or $\{n \in \mathbb{Z} : n \leq 0\}$ with its natural order.

19. (a) Show that b satisfies the definition of $\text{lub}\{x, y, z\}$, i.e., $x \preceq b$, $y \preceq b$, $z \preceq b$, and if $x \preceq c$, $y \preceq c$, $z \preceq c$, then $b \preceq c$.

In detail: Since $\text{lub}\{x, y\} = a$, $x \preceq a$. Similarly $a \preceq b$, so $x \preceq b$ by (T). In the same way $y \preceq b$, and since $\text{lub}\{a, z\} = b$, we have $z \preceq b$.

Now suppose $x \preceq c$, $y \preceq c$ and $z \preceq c$. Then c is an upper bound for x and y, so $a = \text{lub}\{x, y\} \preceq c$. Then c is an upper bound for a and z, so $b = \text{lub}\{a, z\} \preceq c$. Thus $b \preceq c$ for every upper bound c of $\{x, y, z\}$, and we conclude that $b = \text{lub}\{x, y, z\}$.

(b) Show by induction on n that every n-element subset of a lattice has a least upper bound. This is clear for $n = 1$ and true for $n = 2$ by the definition of lattice. Suppose it is true for some $n \geq 2$, and consider a subset $\{a_1, \ldots, a_n, a_{n+1}\}$. By assumption, $\{a_1, \ldots, a_n\}$ has a least upper bound, say a. Then $\text{lub}\{a, a_{n+1}\}$ is the least upper bound for $\{a_1, \ldots, a_{n+1}\}$ by an argument like the one in part (a). Here $a_1, \ldots, a_n$ play the role that x, y played in part (a). By induction, every finite subset has a least upper bound. [This proof is an illustration of the observation that in many cases "2 = finite." That is, what can be done for 2 objects can be done for any finite number.]

(c) Use part (a) and commutativity of $\vee$: $(x \vee y) \vee z = \text{lub}\{x, y, z\} = \text{lub}\{y, z, x\} = (y \vee z) \vee x$ [by part (a) again] $= x \vee (y \vee z)$.

20. $w \vee (x \wedge y) = w \vee v = w$, $(w \vee x) \wedge (w \vee y) = z \wedge z = z$,

$w \wedge (x \vee y) = w \wedge z = w$, $(w \wedge x) \vee (w \wedge y) = v \vee v = v$.

Section 11.2

1. (a) $\{\emptyset, \{1\}, \{1, 4\}, \{1, 4, 3\}, \{1, 4, 3, 5\}, \{1, 4, 3, 5, 2\}\}$ is one.

(b) $5! = 120$.

2. (a) $(1,1)$, $(1,2)$, $(2,2)$, $(2,3)$, $(2,4)$, $(3,4)$, $(4,4)$ is one example.

(b) No. Each step in the chain increases either the first or second coordinate by at least 1, and only $3 + 3 = 6$ increases are possible.

3. (a) $501, 502, \ldots, 1000$. (b) Some examples are:

$\{2, 4, 8, 16, 32, 64, 128, 256, 512\}$, $\{2, 4, 8, 40, 200, 1000\}$, $\{503\}$.

(c) Yes. Think of primes or see Exercise 17.

4. (a) Yes. The poset $\mathscr{C}(S)$ of all chains in S is ordered by $\subseteq$, so a chain greater than the given one would have more members.

(b) Here are Hasse diagrams for some examples

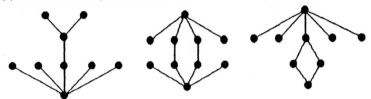

Another example is $\{1, 2, 3, 5, 6, 7, 11, 12, 13\}$ with order relation |.

5. Yes. If $a \preceq b$, then $\text{lub}(a,b) = b$ and $\text{glb}(a,b) = a$.

6. Here is one natural way. For $x \in C_i$ and $y \in C_j$ define $x \preceq y$ if $i < j$ or if $i = j$ and $x \preceq_i y$.

7. (a) Transitivity, for example. If $f \preceq g$ and $g \preceq h$, then $f(t) \preceq g(t)$ and $g(t) \preceq h(t)$ in S, for all t in T. Since $\preceq$ is transitive on S, $f(t) \preceq h(t)$ for all t, so $f \preceq h$ in FUN(T,S).

(b) Suppose $f_m \preceq g$ with $g \in$ FUN(T,S). Then $m = f_m(t) \leq g(t)$ for every $t \in T$. Since m is maximal in S, $m = g(t)$ for all t, so $g = f_m$. Thus f_m is maximal in FUN(T,S).

(c) $f(t) \preceq f(t) \vee g(t) = h(t)$ for all t, so $f \preceq h$. Similarly $g \preceq h$, so h is an upper bound for $\{f, g\}$. Show that if $f \preceq k$ and $g \preceq k$ then $h \preceq k$, so that h is the least upper bound for $\{f, g\}$.

(d) $f < g$ in case $f(t) \preceq g(t)$ for every t, and $f(t) < g(t)$ for at least one t.

8.

	0	1	2	3	4	5	6			
	·	·	·	·	·	·	·			
	·	·	·	·	·	·	·			
	·	·	·	·	·	·	·			
5	∘	∘	∘	∘	∘	∘	∘	·	·	·
4	∘	∘	∘	∘	∘	∘	∘	·	·	·
3	∘	∘	∘	∘	∘	∘	∘	·	·	·
2	•	•	•	•	•	•	∘	·	·	·
1	•	•	•	•	•	•	∘	·	·	·
0	•	•	•	•	•	•	∘	·	·	·
	0	1	2	3	4	5	6			

9.   (a)   (b)   (c)

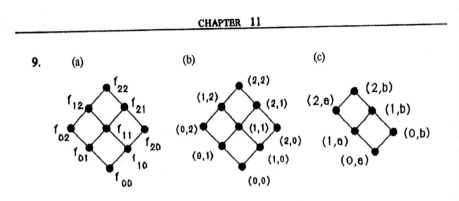

10. No. It need not be antisymmetric or transitive. For example, let $S = T = \mathbb{N}$ with the usual order $\leq$. Then $(0,2) \preceq (1,1)$ since $0 \leq 1$, but $(1,1) \preceq (0,2)$ since $1 \leq 2$. Thus $\preceq$ is not antisymmetric. The string $(2,2) \preceq (3,0) \preceq (1,1)$ shows that transitivity fails too.

11. (a) $(0,0)$, $(0,1)$, $(0,2)$, $(1,0)$, $(1,1)$, $(1,2)$, $(2,0)$, $(2,1)$, $(2,2)$.
   (b) $(0,3)$, $(0,4)$, $(1,3)$, $(1,4)$, $(2,3)$, $(2,4)$.
   (c) $(3,0)$, $(3,1)$, $(3,2)$, $(4,0)$, $(4,1)$, $(4,2)$.

12. If S, say, contains elements s and s' that are not comparable under $\preceq_1$, then (s,t) and (s',t) are not comparable in $S \times T$ for any t in T, so $S \times T$ is not a chain. Thus we may assume S and T are themselves chains. Say $s <_1 s'$ in S and $t <_2 t'$ in T. Then the pairs (s,t') and (s',t) are not comparable in $S \times T$ with the product order.

13. (a) 000, 0010, 010, 10, 1000, 101, 11.
   (b) 10, 11, 000, 010, 101, 0010, 1000.

14. (a) No. If $w \in \Sigma^*$ then $w \preceq^* wu$ for every $u \in \Sigma^*$.
   (b) No, for the same reason.

15. (a) in of the list this order words sentence standard increasing.
   (b) in increasing lexicographic list of order sentence the this words.

16. They are the same only if Σ has exactly one member, since if $a \neq b$ then $ab \prec_L b$ but $b \prec^* ab$.

17. Consider a maximal chain $a_1 < a_2 < \cdots < a_n$ in S. There is no chain $b < a_1 < a_2 < \cdots < a_n$ in S, so there is no b with $b < a_1$. That is, a_1 is minimal. [Finiteness is essential. The chain $(\mathbb{Z}, \leq)$ is a maximal chain in itself.]

176

18. (a) If $(m_1,m_2) \preceq (s,t)$ in the filing order, then $m_1 \preceq_1 s$, so $m_1 = s$ because m_1 is maximal. Since $(s,m_2) \preceq (s,t)$, $m_2 \preceq_2 t$ and so $m_2 = t$ because m_2 is maximal. Thus $(m_1,m_2) = (s,t)$ whenever $(m_1,m_2) \preceq (s,t)$; i.e., (m_1,m_2) is maximal.

(b) No. If (s,t) is maximal in $S \times T$ and $s \preceq_1 s'$ then $(s,t) \preceq (s',t)$, which forces $(s,t) = (s',t)$ and thus $s = s'$. That is, s is maximal in S. If $t \preceq_2 t'$ then $(s,t) \preceq (s,t')$ forces $t = t'$, so t is maximal in T.

(c) Yes. Suppose (s,t) is the largest member of $S \times T$. Then for all $s' \in S$, $t' \in T$, we have $(s',t) \preceq (s,t)$ and $(s,t') \preceq (s,t)$, so $s' \preceq_1 s$ and $t' \preceq_2 t$. Thus s and t are largest members of S and T, respectively.

19. Antisymmetry is immediate. For transitivity consider cases. Suppose
$$(s_1,\ldots,s_n) < (t_1,\ldots,t_n) \quad \text{and} \quad (t_1,\ldots,t_n) < (u_1,\ldots,u_n).$$
If $s_1 \prec t_1$ then $s_1 \prec t_1 \preceq u_1$, so $(s_1,\ldots,s_n) < (u_1,\ldots,u_n)$. If $s_1 = t_1, \ldots, s_{r-1} = t_{r-1}$, $s_r \prec t_r$ and $t_1 = u_1, \ldots, t_{p-1} = u_{p-1}, t_p \prec u_p$ and if $r < p$, then $s_1 = u_1, \ldots, s_{r-1} = u_{r-1}$ and $s_r \prec t_r = u_r$, and again $(s_1,\ldots,s_n) < (u_1,\ldots,u_n)$. The remaining cases are similar.

Section 11.3

1. (a) $A*A = \begin{bmatrix} 1 & 1 & 1 \\ 1 & 1 & 1 \\ 1 & 1 & 1 \end{bmatrix}$. Since $A*A \leq A$ is not true, R is not transitive.

(b) $A * A = A$, so R is transitive.

(c) Not transitive. Note that $A * A = \begin{bmatrix} 1 & 0 & 0 \\ 0 & 1 & 0 \\ 0 & 0 & 1 \end{bmatrix}$.

2. (a)   (b)   (c)

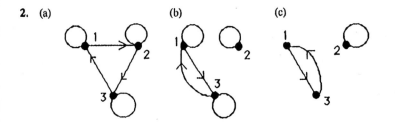

3. (a) The matrix for R is $A = \begin{bmatrix} 0 & 0 & 0 \\ 1 & 0 & 1 \\ 0 & 1 & 0 \end{bmatrix}$. The matrix for R^2 is $A*A = \begin{bmatrix} 0 & 0 & 0 \\ 0 & 1 & 0 \\ 1 & 0 & 1 \end{bmatrix}$.

(b)

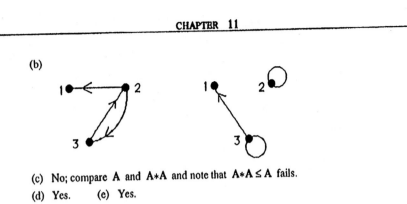

(c) No; compare A and A*A and note that A*A ≤ A fails.

(d) Yes.   (e) Yes.

4. (a) $\begin{bmatrix} 1 & 1 & 1 \\ 0 & 0 & 0 \\ 0 & 1 & 0 \end{bmatrix}$, $\begin{bmatrix} 1 & 1 & 1 \\ 0 & 0 & 0 \\ 0 & 0 & 0 \end{bmatrix}$.

(b)

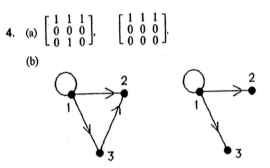

(c) Just observe that $A * A < A$.

(d) $R^2 = R^3 = \cdots = \{(1,1), (1,2), (1,3)\}$.

5. (a) Matrix for R^0 is the identity matrix. Matrix for R^1 is A, of course. Matrix for R^n is A*A for $n \geq 2$, as should be checked by induction.

(b) R is reflexive, but not symmetric or transitive.

6. (a) For $n \geq 2$ the matrix is $\begin{bmatrix} 1 & 1 & 1 \\ 1 & 1 & 1 \\ 1 & 1 & 1 \end{bmatrix}$.

(b) R is symmetric but not reflexive or transitive.

7. (a) $A_f = \begin{bmatrix} 0 & 0 & 1 & 0 \\ 0 & 1 & 0 & 0 \\ 0 & 1 & 0 & 0 \\ 0 & 1 & 0 & 0 \end{bmatrix}$ and $A_g = \begin{bmatrix} 0 & 0 & 0 & 1 \\ 0 & 0 & 1 & 0 \\ 0 & 1 & 0 & 0 \\ 1 & 0 & 0 & 0 \end{bmatrix}$.

(b) They will be different, since the Boolean matrix for $R_f R_g$ is $A_f * A_g = \begin{bmatrix} 0 & 1 & 0 & 0 \\ 0 & 0 & 1 & 0 \\ 0 & 0 & 1 & 0 \\ 0 & 0 & 1 & 0 \end{bmatrix}$;

this is the Boolean matrix for $R_{g \cdot f}$ but not for $R_{f \cdot g}$.

The Boolean matrix for $R_{f \cdot g}$ is $\begin{bmatrix} 0 & 1 & 0 & 0 \\ 0 & 1 & 0 & 0 \\ 0 & 1 & 0 & 0 \\ 0 & 0 & 1 & 0 \end{bmatrix}$.

(c) The Boolean matrix for $R \overset{\leftarrow}{f}$ is $\begin{bmatrix} 0 & 0 & 0 & 0 \\ 0 & 1 & 1 & 1 \\ 1 & 0 & 0 & 0 \\ 0 & 0 & 0 & 0 \end{bmatrix}$.

The Boolean matrix for $R \overset{\leftarrow}{g}$ is $\begin{bmatrix} 0 & 0 & 0 & 1 \\ 0 & 0 & 1 & 0 \\ 0 & 1 & 0 & 0 \\ 1 & 0 & 0 & 0 \end{bmatrix}$. $R \overset{\leftarrow}{g}$ is a function and $R \overset{\leftarrow}{f}$ is not.

8. (a) $\begin{bmatrix} 0 & 0 & 0 & 1 \\ 0 & 0 & 1 & 0 \\ 0 & 1 & 0 & 0 \\ 1 & 0 & 0 & 0 \end{bmatrix}$.

(b) $\begin{bmatrix} 1 & 0 & 1 & 0 \\ 0 & 1 & 0 & 1 \\ 1 & 0 & 1 & 0 \\ 0 & 1 & 0 & 1 \end{bmatrix}$.

(c) $\begin{bmatrix} 1 & 1 & 1 & 1 \\ 0 & 1 & 1 & 1 \\ 0 & 0 & 1 & 1 \\ 0 & 0 & 0 & 1 \end{bmatrix}$.

(d) $\begin{bmatrix} 1 & 1 & 1 & 1 \\ 1 & 1 & 1 & 1 \\ 1 & 1 & 1 & 0 \\ 1 & 1 & 0 & 0 \end{bmatrix}$.

(e) $\begin{bmatrix} 0 & 0 & 0 & 1 \\ 0 & 0 & 0 & 1 \\ 0 & 0 & 0 & 1 \\ 1 & 1 & 1 & 1 \end{bmatrix}$.

9. (a) R_1 satisfies (AR) and (S).
   (b) R_2 satisfies (R), (S) and (T). It's an equivalence relation.
   (c) R_3 satisfies (R), (AS) and (T).
   (d) R_4 satisfies only (S).
   (e) R_5 satisfies only (S).

10. R_3 is the only partial order, and R_2 is the only equivalence relation. R_1, R_4 and R_5 are not transitive, so cannot be partial orders or equivalence relations.

11. (a) True. For each x, $(x,x) \in R_1 \cap R_2$, so $(x,x) \in R_1 R_2$.
    (b) False. The example in part (c) below works. Since $\{(3,2), (2,1)\} \subseteq R_1 R_2$ but $(3,1) \notin R_1 R_2$, this relation is not transitive.
    (c) False. Consider the equivalence relations R_1 and R_2 on $\{1, 2, 3\}$ with Boolean matrices

$$A_1 = \begin{bmatrix} 1 & 1 & 0 \\ 1 & 1 & 0 \\ 0 & 0 & 1 \end{bmatrix} \quad \text{and} \quad A_2 = \begin{bmatrix} 1 & 0 & 0 \\ 0 & 1 & 1 \\ 0 & 1 & 1 \end{bmatrix}.$$

179

Then $A_1 * A_2 = \begin{bmatrix} 1 & 1 & 1 \\ 1 & 1 & 1 \\ 0 & 1 & 1 \end{bmatrix}$. Since $(1,3) \in R_1 R_2$ but $(3,1) \notin R_1 R_2$, this relation is not symmetric.

12. The Boolean matrix is the identity matrix. E is clearly reflexive, symmetric and transitive. "Equality" is the simplest of all equivalence relations.

13. Don't use Boolean matrices; the sets S, T, U might be infinite.
    (a) $R_1 R_3 \cup R_1 R_4 \subseteq R_1 (R_3 \cup R_4)$ by Example 2(a). For the reverse inclusion, consider (s,u) in $R_1 (R_3 \cup R_4)$ and show (s,u) is in $R_1 R_3$ or $R_1 R_4$.
    (b) If $(s,u) \in (R_1 \cap R_2) R_3$, there is a t such that $(s,t) \in R_1 \cap R_2$ and $(t,u) \in R_3$. Since $(s,t) \in R_1$ we have $(s,u) \in R_1 R_3$, and since $(s,t) \in R_2$ we have $(s,u) \in R_2 R_3$. One example where equality fails is given by relations with Boolean matrices
    $$A_1 = \begin{bmatrix} 1 & 0 \\ 0 & 1 \end{bmatrix}, \qquad A_2 = \begin{bmatrix} 0 & 1 \\ 1 & 0 \end{bmatrix} \text{ and } A_3 = \begin{bmatrix} 1 & 1 \\ 1 & 1 \end{bmatrix}.$$
    (c) Show that $R_1 (R_3 \cap R_4) \subseteq R_1 R_3 \cap R_1 R_4$. Equality need not hold. For example, consider R_1, R_3, R_4 with Boolean matrices
    $$A_1 = \begin{bmatrix} 1 & 1 \\ 0 & 0 \end{bmatrix}, \qquad A_3 = \begin{bmatrix} 0 & 0 \\ 0 & 1 \end{bmatrix}, \qquad A_4 = \begin{bmatrix} 0 & 1 \\ 0 & 0 \end{bmatrix}.$$

14. If $(s,u) \in (R_1 R_2)^{\leftarrow}$ then $(u,s) \in R_1 R_2$ so there is a t in T with $(u,t) \in R_1$ and $(t,s) \in R_2$. Then $(s,t) \in R_2^{\leftarrow}$ and $(t,u) \in R_1^{\leftarrow}$, so $(s,u) \in R_2^{\leftarrow} R_1^{\leftarrow}$. This argument is reversible.

15. (R) R is reflexive if and only if $(x,x) \in R$ for every x, if and only if $A[x,x] = 1$ for every x.
    (AR) Similar to the argument for (R), with $(x,x) \notin R$ and $A[x,x] = 0$.
    (S) Follows from $A^T[x,y] = A[y,x]$ for every x, y.
    (AS) R is antisymmetric if and only if $x = y$ whenever $(x,y) \in R$ and $(y,x) \in R$, i.e., whenever $A[x,y] = A^T[x,y] = 1$. Thus R is antisymmetric if and only if all of the off-diagonal entries of $A \wedge A^T$ are 0.
    (T) This follows from Theorem 3 and (a) of the summary.

16. (a) $R_1 \subseteq R_2$ means that $(s,t) \in R_2$ whenever $(s,t) \in R_1$, i.e., that $A_2[s,t] = 1$ whenever $A_1[s,t] = 1$. Since $A_2[s,t] \geq A_1[s,t]$ whenever $A_1[s,t] = 0$, $R_1 \subseteq R_2$ if and only if $A_2[s,t] \geq A_1[s,t]$ for every s, t.
    (b) $(s,t) \in R_1 \cup R_2$ if and only if $(s,t) \in R_1$ or $(s,t) \in R_2$, if and only if $A_1[s,t] = 1$ or $A_2[s,t] = 1$, if and only if $A_1 \vee A_2[s,t] = 1$.

(c) Similar to (b).

17. Given $m \times n$, $n \times p$ and $p \times q$ Boolean matrices A_1, A_2, A_3, they correspond to relations R_1, R_2, R_3 where R_1 is a relation from $\{1, 2, \dots, m\}$ to $\{1, 2, \dots, n\}$, etc. The matrices for $(R_1 R_2) R_3$ and $R_1 (R_2 R_3)$ are $(A_1 * A_2) * A_3$ and $A_1 * (A_2 * A_3)$ by four applications of Theorem 1.

18. (a) Suppose $(s,s') \in RR^{\leftarrow}$. Then there is a t with $(s,t) \in R$ and $(t,s') \in R^{\leftarrow}$. So $(t,s) \in R^{\leftarrow}$ and $(s',t) \in R$; hence $(s',s) \in RR^{\leftarrow}$.
    (b) Replace R by $R^{\leftarrow}$ in part (a) to get $R^{\leftarrow}(R^{\leftarrow})^{\leftarrow}$ symmetric and note that $(R^{\leftarrow})^{\leftarrow} = R$.
    (c) $RR^{\leftarrow}$ is reflexive if and only if each s in S is related to at least one t in T by R.

19. (a) To show that $R \cup E$ is a partial order, show
    (R) $(x,x) \in R \cup E$ for all $x \in S$,
    (AS) $(x,y) \in R \cup E$ and $(y,x) \in R \cup E$ imply $x = y$,
    (T) $(x,y) \in R \cup E$ and $(y,z) \in R \cup E$ imply $(x,z) \in R \cup E$.
    To verify (T), consider cases. The four cases for (T) are: $(x,y) \in R$ and $(y,z) \in R$, $(x,y) \in R$ and $(y,z) \in E$ [so $y = z$], $(x,y) \in E$ and $(y,z) \in R$, $(x,y) \in E$ and $(y,z) \in E$. The last two can be grouped together, since if $(x,y) \in E$ and $(y,z) \in R \cup E$ then $(x,z) = (y,z) \in R \cup E$.
    (b) $R \setminus E$ is antireflexive for every R. If (x,y), $(y,z) \in R \setminus E$, then (x,y), $(y,z) \in R$ so $(x,z) \in R$. Suppose, if possible, that $(x,z) \in E$. Then $x = z$, so (x,y), $(y,x) \in R$, contradicting antisymmetry of R. Hence $(x,z) \in R \setminus E$.

Section 11.4

1. (a) $\begin{bmatrix} 1 & 1 & 0 \\ 0 & 1 & 0 \\ 0 & 0 & 1 \end{bmatrix}$. (b) $\begin{bmatrix} 0 & 1 & 0 \\ 1 & 0 & 0 \\ 0 & 0 & 1 \end{bmatrix}$. (c) $\begin{bmatrix} 1 & 1 & 0 \\ 1 & 1 & 0 \\ 0 & 0 & 1 \end{bmatrix}$.
   (d) $\begin{bmatrix} 1 & 1 & 0 \\ 1 & 1 & 0 \\ 0 & 0 & 1 \end{bmatrix}$. (e) $\begin{bmatrix} 1 & 1 & 0 \\ 1 & 1 & 0 \\ 0 & 0 & 1 \end{bmatrix}$.

2. (a) $\begin{bmatrix} 1 & 1 & 1 \\ 0 & 1 & 1 \\ 0 & 0 & 1 \end{bmatrix}$. (b) $\begin{bmatrix} 0 & 1 & 1 \\ 1 & 0 & 1 \\ 1 & 1 & 0 \end{bmatrix}$. (c) $\begin{bmatrix} 1 & 1 & 1 \\ 1 & 1 & 1 \\ 1 & 1 & 1 \end{bmatrix} = $ (d) = (e).

3. $\{1, 2\}, \{3\}$.

4. $\{1, 2, 3\}$ is the only class.

5. (a) $\begin{bmatrix} 1 & 1 & 0 & 0 & 0 \\ 0 & 1 & 0 & 1 & 0 \\ 0 & 0 & 1 & 0 & 1 \\ 0 & 1 & 0 & 1 & 0 \\ 0 & 0 & 0 & 0 & 1 \end{bmatrix}.$ (b) $\begin{bmatrix} 0 & 1 & 0 & 0 & 0 \\ 1 & 1 & 0 & 1 & 0 \\ 0 & 0 & 0 & 0 & 1 \\ 0 & 1 & 0 & 0 & 0 \\ 0 & 0 & 1 & 0 & 0 \end{bmatrix}.$ (c) $\begin{bmatrix} 1 & 1 & 0 & 0 & 0 \\ 1 & 1 & 0 & 1 & 0 \\ 0 & 0 & 1 & 0 & 1 \\ 0 & 1 & 0 & 1 & 0 \\ 0 & 0 & 1 & 0 & 1 \end{bmatrix}.$

(d) $\begin{bmatrix} 1 & 1 & 0 & 0 & 0 \\ 1 & 1 & 0 & 1 & 0 \\ 0 & 0 & 1 & 0 & 1 \\ 0 & 1 & 0 & 1 & 0 \\ 0 & 0 & 1 & 0 & 1 \end{bmatrix}.$ (e) $\begin{bmatrix} 1 & 1 & 0 & 1 & 0 \\ 1 & 1 & 0 & 1 & 0 \\ 0 & 0 & 1 & 0 & 1 \\ 1 & 1 & 0 & 1 & 0 \\ 0 & 0 & 1 & 0 & 1 \end{bmatrix}.$

6. $\{1, 2, 4\}, \{3, 5\}$.

7. (a) $r(R)$ is the usual order $\leq$.   (b) $sr(R)$ is the universal relation.

    (c) $rs(R)$ is the universal relation on $\mathbb{P}$.   (d) $tsr(R)$ is the universal relation.

    (e) R is already transitive.   (f) $(m,n) \in st(R)$ if and only if $m \neq n$.

8. (a) R is already reflexive, so $r(R) = R$.

    (b) $sr(R) = \{(m,n) \in \mathbb{P} \times \mathbb{P} : m|n \text{ or } n|m\}$.

    (c) $rs(R) = sr(R)$ as in part (b).

    (d) For every $m,n \in \mathbb{P}$, both $(m,1)$ and $(1,n)$ are in $sr(R)$, so (m,n) is in $tsr(R)$. Thus $tsr(R)$ is the universal relation.

    (e) R is already transitive, so $t(R) = R$.

    (f) Same as in part (b), since $st(R) = s(R) = sr(R)$.

9. $(h_1,h_2) \in st(R)$ if $h_1 = h_2$ or if one of h_1, h_2 is the High Hermit. On the other hand, $ts(R)$ is the universal relation on F.O.H.H.

10. (a) If $(x,y) \in \bigcup_{k=1}^{\infty} R_k$ then $(x,y) \in R_{\ell}$ for some ℓ, and since R_{ℓ} is symmetric

$$(y,x) \in R_{\ell} \subseteq \bigcup_{k=1}^{\infty} R_k.$$

    (b) Use induction on n. Assuming that R^n is symmetric, if $(x,y) \in R^{n+1} = R^n R$ then there is a z with $(x,z) \in R^n$ and $(z,y) \in R$. So $(y,z) \in R$ and $(z,x) \in R^n$ by assumption, and thus $(y,x) \in RR^n = R^{n+1}$. Beware the trap of supposing that if R and S are symmetric, then RS must be, too; see Exercise 11(c) of § 11.3.

    (c) Suppose R is symmetric. Then $R = R^{\leftarrow}$, so [by Exercise 12(a) of § 3.1] we have $(r(R))^{\leftarrow} = (R \cup E)^{\leftarrow} = R^{\leftarrow} \cup E^{\leftarrow} = R \cup E = r(R)$. The relation $t(R)$ is symmetric, by parts (b) and (a) with $R_k = R^k$.

11. (a) Since $R \subseteq r(R)$, $t(R) \subseteq tr(R)$. Since $E \subseteq r(R)$, $E \subseteq tr(R)$. Thus $rt(R) = t(R) \cup E \subseteq tr(R)$. For the reverse containment $tr(R) \subseteq rt(R)$ it is enough to show that $r(R) \subseteq rt(R)$

182

and that $rt(R)$ is transitive, since then $rt(R)$ contains the transitive closure of $r(R)$. Now $r(R) = E \cup R \subseteq E \cup t(R) = rt(R)$, and $rt(R)$ is transitive by part (c) of the lemma following Example 4. Thus $tr(R) \subseteq rt(R)$.

(b) Compare $(R \cup E) \cup (R \cup E)^{\leftarrow}$ and $(R \cup R^{\leftarrow}) \cup E$; see Exercise 12 of § 3.1.

12. (a) $(R_1 \cup R_2) \cup E = R_1 \cup E \cup R_2 \cup E.$

   (b) $s(R_1 \cup R_2) = (R_1 \cup R_2) \cup (R_1 \cup R_2)^{\leftarrow} = R_1 \cup R_2 \cup R_1^{\leftarrow} \cup R_2^{\leftarrow}$ [Exercise 12 of § 3.1] $= (R_1 \cup R_1^{\leftarrow}) \cup (R_2 \cup R_2^{\leftarrow}) = s(R_1) \cup s(R_2).$

   (c) Yes. $(R_1 \cap R_2) \cup E = (R_1 \cup E) \cap (R_2 \cup E).$

   (d) No. For example, let R_1 and R_2 have Boolean matrices $A_1 = \begin{bmatrix} 0 & 1 \\ 0 & 0 \end{bmatrix}$ and $A_2 = \begin{bmatrix} 0 & 0 \\ 1 & 0 \end{bmatrix}$, respectively.

13. (a) By Exercise 12(a) and (b) $sr(R_1 \cup R_2) = sr(R_1) \cup sr(R_2) = R_1 \cup R_2$. Thus $tsr(R_1 \cup R_2) = t(R_1 \cup R_2)$. Apply Theorem 3.

   (b) It is $R_1 \cap R_2$ because of Exercise 9 in § 3.1.

14. It was shown in Example 4 that $R = t(R)$ but $s(R) \neq ts(R)$. Hence $st(R) = s(R) \neq ts(R)$. Alternatively,

$$st(A) = \begin{bmatrix} 1 & 1 & 1 \\ 1 & 1 & 0 \\ 1 & 0 & 1 \end{bmatrix} \neq \begin{bmatrix} 1 & 1 & 1 \\ 1 & 1 & 1 \\ 1 & 1 & 1 \end{bmatrix} = ts(A).$$

15. Any relation that contains R will include the pair $(1,1)$, so will not be antireflexive.

16. The relation given by $\begin{bmatrix} 1 & 1 \\ 1 & 0 \end{bmatrix}$ and the one given by $\begin{bmatrix} 1 & 0 \\ 1 & 1 \end{bmatrix}$ are onto relations containing R whose intersection is R. So if there were a smallest onto relation containing R, it would have to be R itself. But R is not an onto relation.

17. (a) The intersection of all relations that contain R and have property p is the smallest such relation.

   (c) (i) fails. $S \times S$ is not antireflexive.

   (d) (ii) fails. See the answer to Exercise 16.

18. Here is one possibility.

{input: n × n Boolean matrix **A**}

{output: matrix t(**A**)}

Set **W** := **A**.

For k = 1 to n do

    for i = 1 to n do

        for j = 1 to n do

            set $W[i,j] := W[i,j] \lor (W[i,k] \land W[k,j])$.

Set t(**A**) := **W**. □

CHAPTER 12

This chapter contains a concrete and friendly introduction to material that could loosely be classified as "abstract algebra." Because of conflicting demands on their time, many of our students will not be able to take an abstract algebra course. For those students, we have tried to give enough of an introduction to the basic algebraic structures and techniques so they can at least follow many of the algebraic arguments they run into in computer science. All evidence we have seen suggests that even more algebra will be required in computer science in the future. We hope to help prepare students who are not majoring in math to view their own subjects through algebraic glasses. For math majors, of course, this first look at algebra acts as an introduction for later more intensive courses.

A superficial listing of terms and examples is no way to give an appreciation for the power of the algebraic approach. We have chosen instead to work out in some detail the application of groups to counting, beginning with a very concrete look at permutation groups. After that we give an introduction to the ideas of abstract group theory. The last section of the chapter looks at commutative rings, with applications to number theory. As noted in the text, §§ 12.5-12.8 can be studied independently of §§ 12.1-12.4 by glossing over the examples that refer back to the earlier sections. We don't think this is a good idea.

Section 12.1 develops cycle notation for permutations and discusses subgroups of S_n, the only group that we know about at this point. Theorem 2 relates the order of a permutation, as a group element, to its cycle decomposition. Look at examples, of course.

Section 12.2 is where the action is. We see the orbits of G itself and also of its cyclic subgroups. Illustrate the concept of orbit by looking at a picture such as those in Examples 1 and 2 and show how to read the orbits from the columns of the tables. The theorem [really the Generalized Class Equation] is not hard to understand, but it is quite different from anything we've seen so far, except perhaps in Chapter 5. Some examples are essential.

It helps to illustrate the one-to-one correspondence between Gs and G/G_s in a concrete case, for instance for E in Figure 2 with $s = x$ and $s = y$. The mechanical calculations involved in permutation group theory are pretty complicated for this audience, even in small cases, so we have presented worked-out answers for the students to refer to. Beware of asking them to give detailed information about other groups that are at all complicated. Symmetry groups of very elementary graphs can get ugly in a hurry if students don't know much group theory.

Section 12.3 continues the combinatorial theme from § 12.2. Theorem 1 comes out of the blue. Students need to see some examples immediately in order to see what it's saying; the proof has a powerful idea in it, if there's time after the examples. Our approach is to carry out the steps in the proof in the setting of Example 1. Don't dwell on the fact that restriction is a homomorphism. This section introduces the graph colorings that we'll study in § 12.4. Theorem 2 looks forbidding, but it's easy to see what it says about Example 6.

The payoff comes in § 12.4 which stops just short of a full-dress account of the Polya-Burnside theory, but gives the flavor of the subject and provides practical tools.

Section 12.5 introduces the concept of an abstract group and gives the elementary facts. Students who have gone through §§ 12.1 and 12.2 will have concrete examples to keep in mind. The ideas behind Lagrange's Theorem [in Theorems 4 and 5] are mildly surprising and are almost as important as the theorem itself.

A good example for geometric intuition is $G = \mathbb{R}^2$ [or $\mathbb{Z}^2$] with coordinatewise addition as operation and the x-axis as a subgroup. The cosets are then just the horizontal lines in the plane. They all "look just like" the subgroup, but they aren't themselves subgroups [for one thing, they don't contain the origin]. A picture is also useful in Example 6.

One reason the theory of groups has been so successful, in contrast to the theories for semigroups and monoids, is that it has a homomorphism theory that works. Section 12.6 develops the basic facts about group homomorphisms–kernels, natural homomorphisms and the Fundamental Homomorphism Theorem. A key fact is that preimages of points in $\varphi(G)$ are cosets of the kernel of φ. Simple examples, for instance those in unassigned exercises, are very helpful.

Students will never have seen anything quite like the fundamental theorem, so they will need help with it. The first difficulty many students have is with the idea of considering subsets of G as themselves elements in the new group G/K. Remind them that the union operation ∪ combines two members of $\mathscr{P}(S)$ [i.e., two subsets of S] to give a set in $\mathscr{P}(S)$. The natural operation on G/K combines two cosets in an algebraic way to give a coset. Example 7 is important in its own right, as well as a good illustration of the methods.

Section 12.7 gives an introduction to semigroups and monoids. It's primarily a discussion of examples that are already very familiar except for the new terminology. Stress that the power of abstraction is that it allows us to see that we've been dealing with the same ideas over and over, though in several different settings.

Section 12.8 introduces a second operation on a set. The main theme is commutative rings, with special emphasis on $\mathbb{Z}$ and on polynomial rings. The treatment is necessarily pretty sketchy.

Our approach to this material in class is to treat it nonchalantly, as the most natural thing in the world, with no surprises. Nevertheless, you will need to spend at least two days on this section. Examples 9 and 10 tie ring theory to some of the world's oldest and newest algorithms.

Let us know how this revised introduction to algebraic structures worked [or didn't] for you and your students.

Section 12.1

1. (a) (1 5 4 2). (b) (1 2 4)(3 5). (c) (2 5)(3 4).
   (d) (1 3 5 2 4).

2. (a) (1 2 4 5). (b) (1 4 2)(3 5). (c) (2 5)(3 4).
   (d) (1 4 2 5 3).

3.

4.

5. (a) (1 4 2). (b) (1 4 3 2). (c) (1 3 4 2).
   (d) (2 4 3). (e) (1 3)(2 4). (f) (1 2).

6. (a) (2 3 6 4). (b) (1 3 2 5 4). (c) (1 4 6)(2 3).
   (d) (1 3 5 2 4). (e) e, the identity. (f) (2 6 3)(4 5).

7. (a) (1 2) itself. (b) (1 6 2 4). (c) (1 6 3).
   (d) (1 3 4 5 2 6).

8. (a) (1 2 4). (b) (1 2 3 4). (c) (1 2 4 3).
   (d) (2 3 4). (e) (1 3)(2 4). (f) (1 2).

9. (b) $(1\ 4)(1\ 3)(1\ 5)(1\ 2)(1\ 7)$.

   (c) $(k_1\ k_2\ \cdots\ k_m) = (k_1\ k_m)(k_1\ k_{m-1})(k_1\ k_{m-2})\ \cdots\ (k_1\ k_3)(k_1\ k_2)$.

   (d) This follows from part (c) and Theorem 1.

10. The m-th proposition is the equation in the answer to Exercise 9(c). This is clear for
    $m = 2$, since 2-cycles are already transpositions. Now
    $$(k_1\ k_2\ \cdots\ k_m\ k_{m+1}) = (k_1\ k_{m+1})(k_1\ k_2\ \cdots\ k_m)$$
    directly, and if the proposition is true for m then this equals
    $$(k_1\ k_{m+1})(k_1\ k_m)(k_1\ k_{m-1})\ \cdots\ (k_1\ k_3)(k_1\ k_2).$$

11. Answers are 1, 1, 2 and 2, respectively. Observe that e has order 1, g^3 has order 2,
    g^2 and g^4 have order 3, while g and g^5 have order 6. You should satisfy yourself
    that these are true statements.

12. Answers are 1, 3, 2 and 0, respectively, since e has order 1, each of $(1\ 2)$, $(1\ 3)$ and
    $(2\ 3)$ has order 2, $(1\ 2\ 3)$ and $(1\ 3\ 2)$ have order 3, and there are no elements of order
    6.

13. (a) $\{e,\ (2\ 5)\}$.

    (b) $\{e,\ (1\ 5\ 2\ 4),\ (1\ 2)(4\ 5),\ (1\ 4\ 2\ 5)\}$.

    (c) $\{e,\ (1\ 6)(2\ 4\ 3),\ (2\ 3\ 4),\ (1\ 6),\ (2\ 4\ 3),\ (1\ 6)(2\ 3\ 4)\}$.

    (d) $\{e,\ (1\ 3\ 5\ 2)(4\ 6),\ (1\ 5)(2\ 3),\ (1\ 2\ 5\ 3)(4\ 6)\}$.

14. (a) 3 since $(1\ 6)(3\ 6) = (1\ 6\ 3)$. Theorem 2 does not apply directly to $(1\ 6)(3\ 6)$ since
    $(1\ 6)$ and $(3\ 6)$ are not *disjoint* cycles.

    (b) 2 since $(1\ 3\ 5)(2\ 5\ 1) = (1\ 2)(3\ 5)$.

    (c) 4 since $(1\ 3)(2\ 6\ 3) = (1\ 3\ 2\ 6)$.

15. (a) e, $(1\ 2)$, $(1\ 2\ 3)$, $(1\ 2\ 3\ 4)$, $(1\ 2\ 3\ 4\ 5)$ and $(1\ 2)(3\ 4\ 5)$, say. $(1\ 2\ 3\ 4\ 5\ 6)$ also
    has order 6, but it's a cycle.

    (b) Clearly cycles have orders 1, 2, 3, 4, 5, or 6. So consider a permutation written as a
    product of two or more disjoint nontrivial cycles; here we regard the 1-cycles as trivial.
    The possible ways to write 6 as a sum with at least two summands greater than 1 are
    $2 + 2 + 1 + 1$, $2 + 2 + 2$, $3 + 2 + 1$, $3 + 3$ and $4 + 2$. In each case the order of the
    corresponding product of cycles is the least common multiple of the orders of the cycles.
    So the possible orders are $\text{lcm}(2,2,1,1) = 2$, $\text{lcm}(2,2,2) = 2$, $\text{lcm}(3,2,1) = 6$, $\text{lcm}(3,3) = 3$ and $\text{lcm}(4,2) = 4$.

16. (a) $(1\ 2)(3\ 4\ 5)$.   (b) $(1\ 2)(3\ 4\ 5\ 6\ 7)$ and $(1\ 2\ 3)(4\ 5\ 6\ 7)$.

17. Yes. $g^i{\circ}g^j = g^{i+j} = g^j{\circ}g^i$ for all powers of g.

18. (a) From Figure 4, we have $g^2 = (1\ 3\ 2)$, $gh = (1\ 2)$ and $hg = (1\ 3)$.

    (b) Of course. $ghg = (gh)g = (1\ 2)(1\ 2\ 3) = (2\ 3) = h$ and $hgh = (hg)h = (1\ 3)(2\ 3) = (1\ 3\ 2) = g^2$.

19. (a) If j is a multiple of m, say $j = qm$, then $g^j = (g^m)^q = e^q = e$. Conversely, suppose $g^j = e$. Use the Division Algorithm to write $j = qm + r$ where $0 \le r < m$. Then as above $g^{qm} = e$ and so $g^r = g^{qm+r} = g^j = e$. Since m is the least positive exponent for g^m to equal to e, r must be 0. So j is a multiple of m.

    (b) It suffices to show that $c_i^{\,j} = e$. If $c_i^{\,j}$ moves some x, then c_i itself must move it, so by disjointness the other c_i's must fix x. Then $x = e(x) = c_1^{\,j} c_2^{\,j} \cdots c_k^{\,j}(x) = c_i^{\,j}(x)$, which is absurd.

    (c) Use the notation established prior to Theorem 2. Suppose $g^j = e$ and $j > 0$. Then $c_1^{\,j} c_2^{\,j} \cdots c_k^{\,j} = e$. By part (b) each $c_i^{\,j} = e$. Apply part (a) k times to conclude that j is a multiple of each m_i. So j is a common multiple of $m_1, m_2, \ldots, m_k$ and hence $j \ge \mathrm{lcm}(m_1, m_2, \ldots, m_k)$.

20. (a)

	e	(2 5)
e	e	(2 5)
(2 5)	(2 5)	e

$\langle(2\ 5)\rangle$

(b)

	e	(1 5 2 4)	(1 2)(4 5)	(1 4 2 5)
e	e	(1 5 2 4)	(1 2)(4 5)	(1 4 2 5)
(1 5 2 4)	(1 5 2 4)	(1 2)(4 5)	(1 4 2 5)	e
(1 2)(4 5)	(1 2)(4 5)	(1 4 2 5)	e	(1 5 2 4)
(1 4 2 5)	(1 4 2 5)	e	(1 5 2 4)	(1 2)(4 5)

$\langle(1\ 5\ 2\ 4)\rangle$

Section 12.2

1. (a) Let g, f be in AUT(D). Then (x, y) is an edge if and only if $(g(x), g(y))$ is an edge. Similarly, $(g(x), g(y))$ is an edge if and only if $(f(g(x)), f(g(y)))$ is an edge. Hence (x, y)

is an edge if and only if $(f(g(x)),f(g(y)))$ is. So $f \circ g$ is in AUT(D). As we noted in § 12.1, we don't need to check that $g \in$ AUT(D) implies $g^{-1} \in$ AUT(D), though this is easy to do directly.

(b) Suppose for example that x is a source but that $g(x)$ is not, for some $g \in$ AUT(D). Then there is some edge $(y,g(x))$. It follows that $(g^{-1}(y),g^{-1}(g(x))) = (g^{-1}(y),x)$ is also an edge. But then x is the terminal vertex of some edge, hence not a source. A similar argument works for sinks.

2. (a)

	e	g	h	gh	f	fh	fg	fgh
e	e	g	h	gh	f	fh	fg	fgh
g	g	e	gh	h	fh	f	fgh	fg
h	h	gh	e	g	fg	fgh	f	fh
gh	gh	h	g	e	fgh	fg	fh	f
f	f	fg	fh	fgh	e	h	g	gh
fh	fh	fgh	f	fg	g	gh	e	h
fg	fg	f	fgh	fh	h	e	gh	g
fgh	fgh	fh	fg	f	gh	g	h	e

(b) No; the table is not symmetric about the diagonal. For example, $gf = fh \neq fg$.

3. $|\text{AUT}(D)| = 4$. $|\text{AUT}(D)p| = |\{p, r\}| = 2$, $|\text{FIX}(p)| = |\{e, f\}| = 2$ and $2 \cdot 2 = 4$; the cases for q, r and s are similar.

4. $|\text{AUT}(E)| = 6$. $|\text{AUT}(E)x| = |\{x\}| = 1$, $|\text{FIX}(x)| = |\text{AUT}(E)| = 6$ and $6 \cdot 1 = 6$. $|\text{AUT}(E)y| = |\{y, z, w\}| = 3$, $|\text{FIX}(y)| = |\{e, h\}| = 2$ and $3 \cdot 2 = 6$; similarly for z and w.

5. (b) and (c)   See Example 2 in § 12.4.

6. They must be divisors of 27 by the Theorem, i.e., 1, 3, 9 or 27.

7. The Theorem shows that each $|Gx_j|$ must be a divisor of $|G| = 2^k$. So each $|Gx_j|$ must be $1, 2, 4, \ldots$ or 2^k. Since $|X|$ is odd, for some j we must have $|Gx_j| = 1$. By the Theorem again, $|G| = |\text{FIX}(x_j)|$ and so $G = \text{FIX}(x_j)$. Hence $g(x_j) = x_j$ for all $g \in G$.

8. (a) 8. Redraw the tree as

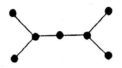

and compare with the graph in Figure 3.

(b) $2 \cdot 8 \cdot 8 = 128$.

(c) By part (b), $|\text{AUT}(T_2)| = 2^7$ and $|T_2| = 15$ is odd. So Exercise 7 applies.

(d) The top vertex is not moved by any automorphism.

9. (a) Here is one possible sequence.

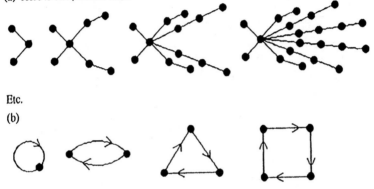

Etc.

(b)

Etc.

10. (a) This part is a special case of part (b) since $\text{FIX}(x) = \text{FIX}(\{x\})$.

(b) If $f(Y) = Y$ and $g(Y) = Y$, then $(f \circ g)(Y) = f(g(Y)) = f(Y) = Y$, so $f \circ g \in \text{FIX}(Y)$. Moreover, $Y = (f^{-1} \circ f)(Y) = f^{-1}(f(Y)) = f^{-1}(Y)$, so $f^{-1} \in \text{FIX}(Y)$.

11. $\text{FIX}(\{w, y\}) = \{e, f\}$ since both $f(w)$ and $f(y)$ belong to $\{w, y\}$. However, $\text{FIX}(w) \cap \text{FIX}(y) = \{e\}$.

12. (a) Suppose that $g \in \bigcap_{y \in Y} \text{FIX}(y)$. For each $y \in Y$, $g(y) = y \in Y$. So g is in $\text{FIX}(Y)$.

13. (a) $\text{AUT}(H)u = \{u, v\}$. Since $u = e(u)$ and $v = f(u)$, the choice $g_1 = e$, $g_2 = f$ will work. Note that $g_1 \circ \text{FIX}(u) = \text{FIX}(u) = \{e, g, h, gh\}$ and that $g_2 \circ \text{FIX}(u) = \{fe, fg, fh, fgh\}$ so that $g_1 \circ \text{FIX}(u) \cup g_2 \circ \text{FIX}(u)$ is indeed $\text{AUT}(H)$.

(b) No. g_1 can be any member of $\{e, g, h, gh\}$, and g_2 any member of $f \circ \{e, g, h, gh\}$.

(c) AUT(H)x = $\{w, x, y, z\}$. Since $w = g(x)$, $x = e(x)$, $y = f(x)$ and $z = fg(x)$, the choice $g_1 = e$, $g_2 = f$, $g_3 = g$, $g_4 = fg$ works. Indeed $g_1 \circ FIX(x) = FIX(x) = \{e, h\}$, $g_2 \circ FIX(x) = \{fe, fh\}$, $g_3 \circ FIX(x) = \{ge, gh\}$ and $g_4 \circ FIX(x) = \{fge, fgh\}$, so that the union of these four disjoint sets is AUT(H).

14. (1) Obviously $\supseteq$ holds. To show $\subseteq$, consider $g \in G$. $g(x)$ has to be in the orbit Gx, so $g(x) = g_j(x)$ for some j. Then $x = g_j^{-1}(g(x))$ and so $g_j^{-1} \circ g \in FIX(x)$. Thus $g = g_j \circ (g_j^{-1} \circ g)$ belongs to $g_j \circ FIX(x)$ and hence to the union.

(2) Consider $i \neq j$ in $\{1, 2, ..., k\}$. Every permutation in $g_i \circ FIX(x)$ maps x to $g_i(x)$ and every permutation in $g_j \circ FIX(x)$ maps x to $g_j(x)$. Since $g_i(x) \neq g_j(x)$, no permutation can belong to both sets.

(3) The function $g \rightarrow g_j \circ g$ maps FIX(x) onto $g_j \circ FIX(x)$, so it suffices to note that this function is one-to-one: $g_j \circ g = g_j \circ h$ implies $g_j^{-1} \circ g_j \circ g = g_j^{-1} \circ g_j \circ h$ implies $g = h$.

15. (a) Say $X = Gx_0$. For any x in X, obviously $Gx \subseteq X$. Consider $y \in X$. Then $y = g_1(x_0)$ for some $g_1 \in G$. Also $x = g_2(x_0)$ for some $g_2 \in G$. So $y = g_1(x_0) = g_1 \circ g_2^{-1}(x) \in Gx$.

(b) If G is finite, so is Gx = X. By the Theorem and part (a) we have $|G| = |Gx| \cdot |FIX(x)| = |X| \cdot |FIX(x)|$ for each $x \in X$.

16. Obviously K acts on X: each g in G is a permutation of G so surely this is true of each g in K. Since $G = \bigcup_{g \in G} Kg$, if Gx is an orbit of g, then $Gx = \bigcup_{g \in G} Kg(x)$, so that Gx is the union of orbits Kg(x) under K.

17. (a) Since $e \in G$ and $e(x) = x$, R contains all (x,x) and is reflexive. Since $g(x) = y \Rightarrow y = g^{-1}(x)$, $(x,y) \in R \Rightarrow (y,x) \in R$, so R is symmetric. For transitivity, note that the conditions $g(x) = y$ and $g'(y) = z$ imply $g' \circ g(x) = z$.

(b) The sets in the partition are the G-orbits in X.

Section 12.3

1. (a) From Example 1, the numbers are 6, 4, 4, 2, 0, 0, 0, 0 and their sum is 16.

(b) From Example 6(c), § 12.2, the answers are 4, 4, 2, 2, 2, 2 and their sum is 16.

(c) The sums agree; we have calculated |S| in the proof of Theorem 1 in two ways, using formulas (1) and (2).

2. From the table in Figure 4 of § 12.2 we see that FIX(e) has 4 elements, FIX(d) and FIX(f) have 2 elements each, and all the other sets FIX(a) are empty. So Theorem 1 tells us that there are $\frac{1}{8}(4 + 2 + 2 + 0 + 0 + 0 + 0 + 0)$ orbits. Indeed, $\{x, y, z, w\}$ is the only orbit.

3. (a) Graph automorphisms can interchange w and y and they can interchange x and z.
(b) FIX(e) = $\{w, x, y, z\}$, FIX(g) = $\{x, z\}$, FIX(h) = $\{w, y\}$ and FIX(gh) = $\emptyset$. So by Theorem 1, there are $\frac{1}{4}(4 + 2 + 2 + 0)$ orbits under G. Indeed, they are $\{w, y\}$ and $\{x, z\}$.

4. (a)

	e_1	e_2	e_3	e_4	e_5
e^*	e_1	e_2	e_3	e_4	e_5
g^*	e_4	e_5	e_3	e_1	e_2
h^*	e_2	e_1	e_3	e_5	e_4
$(gh)^*$	e_5	e_4	e_3	e_2	e_1

(b) FIX(e^*) = $\{e_1, e_2, e_3, e_4, e_5\}$, FIX($g^*$) = FIX($h^*$) = FIX($(gh)^*$) = $\{e_3\}$, so by Theorem 1 there are $\frac{1}{4}(5 + 1 + 1 + 1)$ orbits under G. They are $\{e_3\}$ and $\{e_1, e_2, e_4, e_5\}$.

5. For the orbit $\{w, y\}$, the automorphisms g and gh both move each element. For the orbit $\{x, z\}$, the automorphisms h and gh both move each element.

6. For the orbit $\{e_3\}$, Corollary 2 is true vacuously because $|\{e_3\}| > 1$ fails. For the orbit $\{e_1, e_2, e_4, e_5\}$, the automorphisms g^*, h^* and $(gh)^*$ all move each element.

7. (a)
(b)

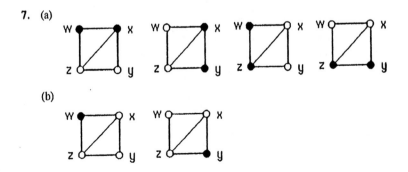

8. (a)   (b)

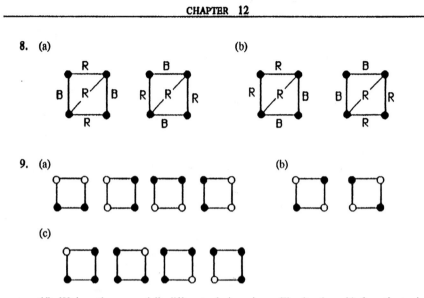

9. (a)   (b)

(c)

(d) We have three essentially different colorings above. The situation with 3 vertices red is similar to part (c). This gives us $4 + 2 + 4 + 4 = 14$ distinct colorings. There are $2^4 = 16$ altogether. The remaining two colorings use all red or all black. So there are 6 essentially different colorings: (a), (b), (c), (c) with 3 red, all red, and all black. In the next section we will have a *much* better method for solving this and much harder problems. See Example 2, § 12.4.

10. Permutations in G^* all take $\{w, x\}$ to $\{w, x\}$ or else to $\{y, z\}$. So any permutations that "break up" the couple $\{w, x\}$ will work. For example, any permutations that take w to w and x to y or z will do.

11. h takes each element of $\{w, x, u, v\}$ to itself, and so h^* takes each two-element subset of this set to itself. There are $\binom{4}{2} = 6$ such sets. The only other two-element set mapped to itself is $\{y, z\}$, so $FIX_T(h^*)$ has 7 elements.

12. (a) $|G| = 8 = 2^3$ and $|E| = 5$ is odd. So Exercise 7, § 12.2, shows that some edge must be fixed by all members of G.
(b) The edge e_3 is fixed by all members of G.

13. (a) $|G| = 2^2$ and $|E| = 5$ is odd, so Exercise 7, § 12.2, shows that some edge must be fixed by all members of G.
(b) The edge e_3 is fixed by all members of G.

14. This g^* doesn't have the fundamental property (∗). Indeed, with this definition $g \to g^*$ is a so-called "anti-homomorphism": $(g \circ h)^* = h^* \circ g^*$.

15. (a) $G = \{e, g\}$ where e is the identity permutation of $V = \{u, v\}$ and $g(u) = v$, $g(v) = u$. Both e^* and g^* are the identity on the one-element edge set E.

(b) Since $g^* = h^*$ implies $(h^{-1} \circ g)^* = e^*$, it suffices to show that $g^* = e^*$ implies $g = e$. [For then $g^* = h^*$ implies $(h^{-1} \circ g)^* = e^*$ implies $h^{-1} \circ g = e$ implies $g = h$.] So assume $g^* = e^*$. Since $g^*(\{u, v\}) = \{g(u), g(v)\} = \{u, v\}$, all that g^* can do to an edge is leave it alone or turn it end for end. If $g \neq e$, then g^* must turn some edge $\{u, v\}$ end for end. Then it has to take any *other* edge $\{u, w\}$ attached to u and attach it to v, while at the same time leaving it alone or switching it end for end. Think about this. We would have $\{u, w\} \to \{v, t\}$ for some t, but also $\{u, w\} \to \{u, w\}$, so $v = w$, $t = u$ and $\{u, w\} = \{u, v\}$. Thus $\{u, w\}$ could not really be an edge different from $\{u, v\}$. So there are no other edges attached to u or, by symmetry, to v. But then, since H is connected, $\{u, v\}$ is the only edge there is, and H is the graph of part (a).

16. Always $|\text{FIX}_X(e)| = n$ and by assumption $|\text{FIX}_X(g)| \geq 1$ for all $g \in G$. So by Theorem 1 the number of orbits is

$$|G|^{-1}\left\{|\text{FIX}_X(e)| + \sum_{g \neq e} |\text{FIX}_X(g)|\right\} \geq |G|^{-1}\{n + |G| - 1\} = 1 + \frac{n-1}{|G|}.$$

Section 12.4

1. (a) 4. (b) None do.

(c) Apply Theorem 2, noting $m(e) = 5$, $m(f) = 4$, $|G| = 2$. Answer is $(k^5 + k^4)/2$.

2. (a) $|\text{FIX}_X(e)| = 5$, $|\text{FIX}_X(f)| = 3$, $|G| = 2$, so the average in Theorem 1, § 12.3, is $\frac{1}{2}(5 + 3) = 4$. This is indeed the number of orbits of G.

(b) Here is how the group acts on edges:

	$\{v, w\}$	$\{w, x\}$	$\{x, y\}$	$\{x, z\}$
e	$\{v, w\}$	$\{w, x\}$	$\{x, y\}$	$\{x, z\}$
f	$\{v, w\}$	$\{w, x\}$	$\{x, z\}$	$\{x, y\}$

Now $|\text{FIX}_X(e)| = 4$, $|\text{FIX}_X(f)| = 2$ and the number of orbits is indeed $\frac{1}{2}(4 + 2) = 3$.

(c) For the types a, b, c, d and e of rotations g, the corresponding values of $|\text{FIX}_X(g)|$ are 0, 2, 2, 0 and 6. The average in Theorem 1, § 12.3, is

$$\frac{1}{24}(6 \cdot 0 + 6 \cdot 2 + 3 \cdot 2 + 8 \cdot 0 + 1 \cdot 6) = 1.$$

G has only one orbit.

3. (a) Here

	w	x	y	z
e	w	x	y	z
a	z	y	x	w

.

$m(e) = 4$, $m(a) = 2$, $|G| = 2$. Theorem 2 gives $C(k) = (k^4 + k^2)/2$.

(b) As in Example 4, the answer is

$$C(4) - 4 \cdot C(3) + 6 \cdot C(2) - 4 \cdot C(1) = 136 - 4 \cdot 45 + 6 \cdot 10 - 4 \cdot 1 = 12.$$

Alternatively, consider the 24 permutations of the labels and note that two permutations give equivalent labels only if they are reverses of each other. E.g., the labels w x y z and z y x w are reverses of each other.

4. By direct calculation check that $g \circ c_1 = c_2 \circ g$. Note that $g \circ g = e$. Hence $g \circ c_1 \circ g = c_2 \circ g \circ g = c_2 \circ e = c_2$.

5. (a) Apply Theorem 2, noting $m(e) = 4$, $m(r) = m(r^2) = 2$, $m(f) = m(g) = m(h) = 3$.
Answer is $C(k) = (k^4 + 2k^2 + 3k^3)/6$.

(b) We have

	{x,w}	{y,w}	{z,w}
e	{x,w}	{y,w}	{z,w}
r	{y,w}	{z,w}	{x,w}
r^2	{z,w}	{x,w}	{y,w}
f	{x,w}	{z,w}	{y,w}
g	{z,w}	{y,w}	{x,w}
h	{y,w}	{x,w}	{z,w}

so $m(e) = 3$, $m(r) = m(r^2) = 1$, $m(f) = m(g) = m(h) = 2$. Apply Theorem 2 to get $C(k) = (k^3 + 2k + 3k^2)/6$.

6. (a) $C(3) - 3 \cdot C(2) + 3 \cdot C(1) = 30 - 3 \cdot 8 + 3 \cdot 1 = 9$.

(b)

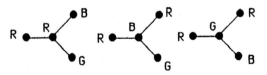

gives the 3 colorings with 2 **red** vertices. Likewise there are 3 colorings with 2 blue vertices and 3 with 2 green vertices.

196

(c) $C(4) - 4 \cdot C(3) + 6 \cdot C(2) - 4 \cdot C(1) = 80 - 4 \cdot 30 + 6 \cdot 8 - 4 \cdot 1 = 4$. The 4 colorings depend only on what color the central vertex is.

7. (a) Using Figure 4(a),
$$C(4) - 4 \cdot C(3) + 6 \cdot C(2) - 4 \cdot C(1) = 55 - 4 \cdot 21 + 6 \cdot 6 - 4 \cdot 1 = 3.$$

(b) Here they are.

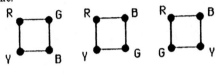

8.

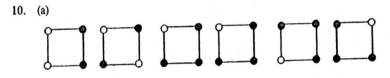

	v	w	x	y	z
e	v	w	x	y	z
type r	various rotations (4 of them)				
f (flip)	v	z	y	x	w
type f∘r	various rotations followed by a flip (4 of them)				

Now $m(e) = 5$, $m(r) = 1$ for each rotation r, $m(f) = 3$ and $m(f \circ r) = 3$ for all automorphisms of type f∘r. Consequently $C(k) = (k^5 + 5k_3 + 4k)/10$.

9. (a) $C(3) - 3 \cdot C(2) + 3 \cdot C(1) = 21 - 3 \cdot 6 + 3 \cdot 1 = 6$.   (b) $C(2) - 2 \cdot C(1) = 6 - 2 = 4$.
(c) $C(5) - 5 \cdot C(4) + 10 \cdot C(3) - 10 \cdot C(2) + 5 \cdot C(1) = 120 - 5 \cdot 55 + 10 \cdot 21 - 10 \cdot 6 + 5 = 0$, of course.

10. (a)

(b) See Figure 4(b). Ignore the graphs with all black and all red vertices.

11. (a) From Figure 4(b), § 12.3, we find: orbits under $\langle e \rangle$ are $\{w\}, \{x\}, \{y\}, \{z\}$; orbits under $\langle g \rangle$ are $\{w, y\}, \{x\}, \{z\}$; orbits under $\langle h \rangle$ are $\{w\}, \{x, z\}, \{y\}$; and orbits under $\langle gh \rangle$ are $\{w, y\}, \{x, z\}$. So $m(e) = 4$, $m(g) = m(h) = 3$ and $m(gh) = 2$.

(b) $C(k) = \frac{1}{4}(k^4 + 2k^3 + k^2)$ using Theorem 2 and part (a).

(c) $C(2) = \frac{1}{4}(16 + 16 + 4) = 9$.

12. (a) From the answer to Exercise 4, § 12.3, we find: e^* has 5 orbits; orbits of g^* are $\{e_1, e_4\}, \{e_2, e_5\}, \{e_3\}$; orbits of h^* are $\{e_1, e_2\}, \{e_3\}, \{e_4, e_5\}$; and orbits of $(gh)^*$ are $\{e_1, e_5\}, \{e_2, e_4\}, \{e_3\}$. So $m(e^*) = 5$ and $m(g^*) = m(h^*) = m((gh)^*) = 3$.

(b) $C(k) = \frac{1}{4}(k^5 + 3k^3)$.

(c) $C(2) - 2 \cdot C(1) = 14 - 2 = 12$.

13. Following the suggestion:

Type	Number of that type	$m(g)$ when group acts on vertices	$m(g)$ when group acts on edges
a	6	4	7
b	6	2	3
c	3	4	6
d	8	4	4
e	1	8	12

(a) $C(k) = (k^8 + 17k^4 + 6k^2)/24$. For example, there are $6 + 3 + 8 = 17$ rotations with $m(g) = 4$.

(b) $C(k) = (k^{12} + 6k^7 + 3k^6 + 8k^4 + 6k^3)/24$.

14. (a) $3 \cdot [C(2) - 2 \cdot C(1)] = 3 \cdot [10 - 2 \cdot 1] = 24$.

(b) Blue faces are adjacent or not. Answer = 2.

(c) The other faces are adjacent or not; and they are either both blue, both green or one each. Answer = 6.

(d) No. It gets complicated fast.

15. (a) In this case, functions numbered n and $15 - n$ in Figure 8 are regarded as equivalent. So there are 8 equivalence classes, which correspond to 8 essentially different 2-input logical circuits.

(b) Now, as noted in the discussion of Figure 8, 2 and 4 are equivalent, as are 3 and 5. So distinct classes are now $\{0, 15\}$, $\{1, 14\}$, $\{2, 4, 11, 13\}$, $\{3, 5, 10, 12\}$, $\{6, 9\}$, $\{7, 8\}$, a total of 6.

Section 12.5

1. (a) $\mathbb{Z}$. (b) $\{0\}$. (c) $\mathbb{Z}$. (d) $\mathbb{Z}$. (e) $\mathbb{Z}$.
   (f) $6\mathbb{Z} \cap 9\mathbb{Z} = 18\mathbb{Z}$.

2. All subgroups of $(\mathbb{Z}, +)$ are cyclic; see Theorem 3. In fact, $n\mathbb{Z} = \langle n \rangle$ for all $n \in \mathbb{N}$.

3. (a) $3\mathbb{Z}$.
   (b) Not a subgroup. For example, 5 is in the set but -5 is not.
   (c) Not a subgroup; 2 and 4 are in the subset, but $2 + 4$ is not.
   (d) $4\mathbb{Z}$.   (e) $\mathbb{N} \cup (-\mathbb{N}) = \mathbb{Z} = 1 \cdot \mathbb{Z}$.

4. All but $\mathbb{N}$ are subgroups of $\mathbb{R}$. [1 belongs to $\mathbb{N}$ but -1 doesn't.] For the other sets S, check that they are closed under addition $[x, y \in S \Rightarrow x + y \in S]$ and negation $[x \in S \Rightarrow -x \in S]$.

5. $h \cdot g = k \cdot g$ implies $h = h \cdot e = h \cdot (g \cdot g^{-1}) = (h \cdot g) \cdot g^{-1} = (k \cdot g) \cdot g^{-1} = k \cdot (g \cdot g^{-1}) = k \cdot e = k$. Also $(h^{-1} \cdot g^{-1})(g \cdot h) = h^{-1} \cdot (g^{-1} \cdot (g \cdot h)) = h^{-1} \cdot ((g^{-1} \cdot g) \cdot h) = h^{-1} \cdot (e \cdot h) = h^{-1} \cdot h = e$.

6. Replace each product $\cdot$ by $+$, and replace each inverse $^{-1}$ by a negation. Thus (a) asserts that $g + h = g + k$ implies $h = k$, and (b) reads $-(g + h) = (-g) + (-h)$. Here's a proof of (a): If $g + h = g + k$, then $h = 0 + h = (-g + g) + h = (-g) + (g + h) = (-g) + (g + k) = (-g + g) + k = 0 + k = k$. The changes for the proof of (b) are similar. There is only one case because G is commutative.

7. (a) We know 1 is a generator. So is 2, since $2 +_5 2 = 4$, $2 +_5 2 +_5 2 = 1$ and $2 +_5 2 +_5 2 +_5 2 = 3$. Similarly, 3 and 4 are generators.
   (b) In this case 1 and 5 are the only generators. Observe that $\langle 2 \rangle = \{0, 2, 4\}$, $\langle 3 \rangle = \{0, 3\}$ and $\langle 4 \rangle = \{0, 4, 2\}$.

8. (a) $\{0\}$.   (b) Yes; it's generated by 0.

9. (a) $\varphi(n) = n + 3$ defines one.   (b) $\psi(n) = n - 1$ is another.

10. $\mathbb{Z} = \bigcup_{r=0}^{4} (5\mathbb{Z} + r)$.

11. The sets in (b) and (c) both contain the permutation $(1\ 2\ 3)$. Neither of the sets contains $(1\ 2\ 3)(1\ 2\ 3) = (1\ 3\ 2)$. The subsets in (a) and (d) are the subgroups $\text{FIX}_G(4)$ and $\text{FIX}_G(\{1, 2\})$; see Exercise 10 of § 12.2.

12. (a) By Lagrange's theorem, the size of any subgroup of S_4 must divide $|S_4| = 24$. Since a subgroup generated by a 13-element set must have at least 13 elements, it follows that it must have 24 elements. I.e., it must be S_4.

   (b) No. Cyclic groups are commutative. $(S_4, \circ)$ is not.

13. (a) Apply Theorem 1(b) twice to get $(g_1 \cdot g_2 \cdot g_3)^{-1} = ((g_1 \cdot g_2) \cdot g_3)^{-1} = g_3^{-1} \cdot (g_1 \cdot g_2)^{-1} = g_3^{-1} \cdot (g_2^{-1} \cdot g_1^{-1})$.

   (b) An easy induction shows that $(g_1 \cdot g_2 \cdots g_n)^{-1} = g_n^{-1} \cdots g_2^{-1} \cdot g_1^{-1}$. The inductive step uses $(g_1 \cdot g_2 \cdots g_n \cdot g_{n+1})^{-1} = g_{n+1}^{-1} \cdot (g_1 \cdot g_2 \cdots g_n)^{-1}$.

14. (a) By definition, $h^{-n} = (h^{-1})^n$ for $n > 0$. For $k < 0$ apply this to $n = -k$ and $h = g$ to get $g^k = (g^{-1})^{-k}$. For $k > 0$ apply it to $n = k$ and $h = g^{-1}$ to obtain $(g^{-1})^{-k} = ((g^{-1})^{-1})^k = g^k$.

   (b) If $m \in \mathbb{N}$, then $g^{m+1} = g^m g$ is the definition of g^{m+1}. If $m = -k$ with $k \in \mathbb{P}$, then $g^m g = g^{-k} g = (g^{-1})^k g = (g^{-1})^{k-1} g^{-1} g = (g^{-1})^{k-1} = g^{-(k-1)} = g^{-k+1} = g^{m+1}$.

   (c) Let $p(n)$ be "$g^m g^n = g^{m+n}$ for all $g \in G$ and $m \in \mathbb{Z}$." Then $p(0)$ is clear, and (b) shows $p(1)$. Use induction on n. For the inductive step
   $$g^m g^{n+1} = g^m g^{1+n} = g^m g^1 g^n \text{ [by } p(n)] = g^{m+1} g^n \text{ [by } p(1)] = g^{m+1+n} \text{ [by } p(n)].$$

   (d) This follows from (c) if $n \in \mathbb{N}$. Suppose $n = -k$ with $k \in \mathbb{P}$. Then $g^m g^{-k} = (g^{-1})^{-m} (g^{-1})^k$ [by (a)] $= (g^{-1})^{-m+k}$ [by (c)] $= g^{m-k}$ [by (a)].

15. (a) If g and h belong to the intersection, then they both belong to each subgroup in the collection, so their product $g \cdot h$ does too. Since each subgroup contains the identity, so does the intersection. Every member of the intersection belongs to each subgroup, so its inverse does too. Thus its inverse also belongs to the intersection. Hence the intersection is closed under products and inverses and contains the identity.

   (b) Any example in which neither H nor K contains the other will work. For example, if $G = (\mathbb{Z}, +)$, let $H = 2\mathbb{Z}$ and $K = 3\mathbb{Z}$.

16. (a) R is generated by the rotation ρ through $72°$. In fact, ρ^2 is the rotation through $144°$, etc., and ρ^5 is the identity map, i.e., the rotation through $0°$. R is also generated by ρ^2, by ρ^3 and by ρ^4.

(b) The identity has order 1 and ρ has order 5. The map g in Example 8(d) has order 2. If some map f in G had order 10, then $|\langle f \rangle|$ would be 10, so $\langle f \rangle$ would be G. But G is not commutative [for example, $h \circ \rho \neq \rho \circ h$], so G is not cyclic; contradiction.

17. (a) $6 = 3!$.   (b) $\text{FIX}_G(1) = \{e, (2\ 3)\}$.
   (c) $\text{FIX}_G(1) \circ (1\ 2\ 3) = \{(1\ 2\ 3), (1\ 3)\}$.
   (d) $(1\ 2\ 3) \circ \text{FIX}_G(1) = \{(1\ 2\ 3), (1\ 2)\} \neq \{(1\ 2\ 3), (1\ 3)\}$.
   (e) It contains $(1\ 2\ 3)$ so the only left coset it *could* be is $(1\ 2\ 3) \circ \text{FIX}_G(1)$.
   Apply part (d).
   (f) 3.

18. (a) It is isomorphic to $(\mathbb{Z}(2),+)$ where $e \to 0$ and $a \to 1$.
   (b) If G were a group, $|\{e, a\}|$ would divide $|G|$, i.e., 2 would divide 5.

19. (a) $\langle a \rangle = \{e, a, b\}$.
   (b) $\langle a \rangle \bullet c = \{e \bullet c, a \bullet c, b \bullet c\} = \{c, d, f\}$ while $c \bullet \langle a \rangle = \{c \bullet e, c \bullet a, c \bullet b\} = \{c, f, d\}$.
   (c) $\langle c \rangle, \langle d \rangle, \langle f \rangle$.
   (d) $|G|/|\langle d \rangle| = 6/2 = 3$. Also, see part (e).
   (e) $\{e, d\}, \{a, c\}, \{b, f\}$.

20. (a) $\langle a \rangle = \{e, a, b\}$.   (b) $\langle a \rangle \bullet c = \{c, d, f\} = c \bullet \langle a \rangle$.
   (c) $\langle d \rangle$.   (d) 3.   (e) $\{e, d\}, \{a, f\}, \{b, c\}$.

21. $g \cdot H$ contains $g \cdot e = g$. So if $g \cdot H = H$, then $g \in H$. If $g \in H$, then H and $g \cdot H$ both contain g, so are not disjoint. By Theorem 4, $g \cdot H = H$ in this case.

22. If $g \in H$, then $g \cdot H = H = H \cdot g$. If $g \notin H$, then $g \cdot H = G \setminus H$ since $\{H, g \cdot H\}$ partitions G; similarly $H \cdot g = G \setminus H$. Note that this shows that the results in Exercises 19(b) and 20(b) hold automatically, since $|G| = 6$ and $|\langle a \rangle| = 3$.

23. (a) For $h \in H$, $(g \cdot h)^{-1} = h^{-1} \cdot g^{-1} \in H \cdot g^{-1}$, so $\{f^{-1} : f \in g \cdot H\} \subseteq H \cdot g^{-1}$. Moreover, $h \cdot g^{-1} = (g \cdot h^{-1})^{-1}$ is in $\{f^{-1} : f \in g \cdot H\}$ for $h \in H$ so $H \cdot g^{-1} \subseteq \{f^{-1} : f \in g \cdot H\}$.
   (b) $g \cdot H \to H \cdot g^{-1}$. Show this is one-to-one.

24. (a) To see that h^* itself is one-to-one, use a cancellation law: $h^*(g_1) = h^*(g_2) \Rightarrow g_1 \cdot h^{-1} = g_2 \cdot h^{-1} \Rightarrow g_1 = g_2$. Each h^* maps G onto G because given $g \in G$ we have $h^*(gh) = g$. So $h \to h^*$ maps H into PERM(G). Also

$$(h_1 \cdot h_2)^*(g) = g \cdot (h_1 \cdot h_2)^{-1} = g \cdot h_2^{-1} h_1^{-1} = h_2^*(g) \cdot h_1^{-1} = h_1^*(h_2^*(g)),$$

so that $(h_1 \cdot h_2)^* = h_1^* \circ h_2^*$. Just as in § 12.3 [see after Example 4 there], this last condition is required to say that H "acts" on G.

(b) If $h_1^* = h_2^*$, then $h_1^*(e) = h_2^*(e)$, and so $h_1^{-1} = h_2^{-1}$. Hence $h_1 = h_2$.

25. (a) (R) Note that $g^{-1} \cdot g = e \in H$.

(S) Note that $g_1^{-1} \cdot g_2 = (g_2^{-1} \cdot g_1)^{-1}$.

(T) Note that $g_3^{-1} \cdot g_1 = (g_3^{-1} \cdot g_2) \cdot (g_2^{-1} \cdot g_1)$.

(b) Show $g_1 \cdot H = \{g \in G : g \sim g_1\}$ as follows. If $h \in H$, then $g_1^{-1} \cdot (g_1 \cdot h) = h \in H$, so $g_1 \cdot h \sim g_1$; thus $g_1 \cdot H \subseteq \{g \in G : g \sim g_1\}$. If $g \sim g_1$, then $g_1^{-1} \cdot g \in H$, so $g = g_1 \cdot (g_1^{-1} \cdot g) \in g_1 \cdot H$; thus $\{g \in G : g \sim g_1\} \subseteq g_1 \cdot H$.

26. $K = g \cdot H \cdot g^{-1}$ contains the identity $e = g \cdot e \cdot g^{-1}$. If $g \cdot h_1 \cdot g^{-1}$ and $g \cdot h_2 \cdot g^{-1}$ are in K, so is their product $(g \cdot h_1 \cdot g^{-1}) \cdot (g \cdot h_2 \cdot g^{-1}) = g \cdot (h_1 \cdot h_2) \cdot g^{-1}$ since $h_1 \cdot h_2 \in H$. If $g \cdot h \cdot g^{-1}$ is in K, so is its inverse $g \cdot h^{-1} \cdot g^{-1}$.

Section 12.6

1. (a), (c), (e).

2. Only (a) and (c) are homomorphisms. In (b), note that $2 = \varphi(1 + 1) \neq \varphi(1) \cdot \varphi(1) = 1$. In (d), $4 = \varphi(1 + 1) \neq \varphi(1) \cdot \varphi(1) = 1$. In (e), $8 = \varphi(1 + 1) \neq \varphi(1) \cdot \varphi(1) = 16$.

3. (a) Not an isomorphism, since it doesn't map $\mathbb{Z}$ *onto* $\mathbb{Z}$.

(b) and (d) are not even homomorphisms.

(c) Is an isomorphism, being a one-to-one and onto homomorphism.

(e) Not an isomorphism, since $\varphi(n) = 3n$ does not map $\mathbb{Z}$ *onto* $\mathbb{Z}$.

4. The mappings in (a) and (b) are not onto $\mathbb{Z}$ and $\mathbb{R} \setminus \{0\}$, respectively. The logarithm homomorphism in (c) is an isomorphism; its inverse is the mapping exp with $\exp(x) = 2^x$.

5. (a) Associativity follows from associativity in $\mathbb{R}$: $[f + (g + h)](x) = f(x) + (g + h)(x) = f(x) + [g(x) + h(x)] = [f(x) + g(x)] + h(x) = (f + g)(x) + h(x) = [(f + g) + h](x)$ for all $x \in \mathbb{R}$. The zero function $\mathbf{0}$, where $\mathbf{0}(x) = 0$ for all $x \in \mathbb{R}$, is the additive identity for F. The additive inverses are just the negatives of the functions.

(b) Yes. $(f + g)(x) = f(x) + g(x) = g(x) + f(x) = (g + f)(x)$ $\forall x \in \mathbb{R}$.

(c) $\varphi(f + g) = (f + g)(73) = f(73) + g(73) = \varphi(f) + \varphi(g)$.

6. Associativity: $(g \cdot K) * [(h \cdot K) * (k \cdot K)] = (g \cdot K) * (h \cdot k \cdot K) = g \cdot (h \cdot k) \cdot K = (g \cdot h) \cdot k \cdot K =$ $(g \cdot h \cdot K) * (k \cdot K) = [(g \cdot K) * (h \cdot K)] * (k \cdot K)$. Also $(g \cdot K) * K = (g \cdot K) * (e \cdot K) = g \cdot e \cdot K = g \cdot K$; similarly $K * (g \cdot K) = g \cdot K$. Finally $(g \cdot K) * (g^{-1} \cdot K) = g \cdot g^{-1} \cdot K = e \cdot K = K$.

7. (a) $\{0\}$.   (b) $\mathbb{Z}$.   (c) $5\mathbb{Z} = \{5n : n \in \mathbb{Z}\}$.   (d) $\{0\}$.

8. (a) $\{73\}$.   (b) $\mathbb{Z}$.   (c) $5\mathbb{Z} + 3$.   (d) $\{73\}$.

9. (a) 4.   (b) 4.   (c) 3.

10. (a) The mapping φ is certainly well-defined. Since $m +_{pq} n = (m + n) \bmod pq$, $m +_{pq} n$ and $m + n$ differ by a multiple of pq; hence they differ by a multiple of p. Thus $\varphi(m +_{pq} n) = (m +_{pq} n) \bmod p = (m + n) \bmod p = (m \bmod p) +_p (n \bmod p) = \varphi(m) +_p \varphi(n)$.

   (b) $\mathbb{Z}(p)$.

   (c) $\{n \in \mathbb{Z}(pq) : p | n\} = \{kp : 0 \le k \le q - 1\} = \{0, p, 2p, \ldots, (q-1)p\}$.

11. (a) The identity is (e_G, e_H), where e_G and e_H are the respective identities of G and H, and $(g, h)^{-1} = (g^{-1}, h^{-1})$.

   (b) $\pi((g_1, h_1) \square (g_2, h_2)) = \pi(g_1 \cdot g_2, h_1 \bullet h_2) = g_1 \cdot g_2 = \pi(g_1, h_1) \cdot \pi(g_2, h_2)$.

   (c) $\{(e_G, h) : h \in H\}$.

   (d) $\{(e_G, h) : h \in H\}$. Part (c) shows that this subgroup is normal.

12. (a) Verification is the same in both coordinates.

   (b) $\varphi(\mathbb{Z}) = \{(0,0), (1,1)\}$, and $(1,1) \square (1,1) = (0,0)$, the identity.

   (c) The kernel of φ is $2\mathbb{Z}$.

   (d) $|\mathbb{Z}/2\mathbb{Z}| = 2$ and $|\mathbb{Z}(2) \times \mathbb{Z}(2)| = 4$, so the groups cannot be isomorphic. $\varphi(\mathbb{Z}) \ne \mathbb{Z}(2) \times \mathbb{Z}(2)$. Compare Example 8(b).

13. (b) $\begin{bmatrix} 0 & 1 \\ 1 & 0 \end{bmatrix} \cdot H = \left\{ \begin{bmatrix} 0 & 1 \\ 1 & x \end{bmatrix} : x \in \mathbb{R} \right\}$ while $H \cdot \begin{bmatrix} 0 & 1 \\ 1 & 0 \end{bmatrix} = \left\{ \begin{bmatrix} x & 1 \\ 1 & 0 \end{bmatrix} : x \in \mathbb{R} \right\}$.

   (c) $\begin{bmatrix} y & z \\ 0 & 1/y \end{bmatrix} \cdot \begin{bmatrix} 1 & x \\ 0 & 1 \end{bmatrix} \cdot \begin{bmatrix} 1/y & -z \\ 0 & y \end{bmatrix} = \begin{bmatrix} 1 & xy^2 \\ 0 & 1 \end{bmatrix}$ is in H.

   (d) Kernel of φ is H.

   (e) Use the result of part (d) and the Fundamental Homomorphism Theorem.

14. (a) This follows from Theorem 1(b), § 12.5.

   (b) The mapping $g \to g^{-1}$ is an isomorphism if and only if G is commutative.

(c) Say $\psi_2 : (G, \cdot) \to (G_0, \bullet)$ and $\psi_1 : (G_0, \bullet) \to (G_1, \square)$. Then $\psi_1 \circ \psi_2$ maps $(G, \cdot)$ into $(G_1, \square)$ and

$$\psi_1 \circ \psi_2(g \cdot h) = \psi_1(\psi_2(g \cdot h)) = \psi_1(\psi_2(h) \bullet \psi_2(g)) = \psi_1(\psi_2(g)) \square \psi_1(\psi_2(h)).$$

15. The pre-image of $\varphi(g)$ is $g \cdot K$ where K is the kernel of φ. By assumption $|g \cdot K| = 1$. So $|K| = 1$ and φ is one-to-one as noted in the corollary to Theorem 1.

16. $J \cap K$ is a subgroup by Exercise 15(a), § 12.5. As noted at the end of this section, for normality it is enough to show $g \cdot (J \cap K) \cdot g^{-1} \subseteq J \cap K$. But $h \in J \cap K$ implies $g \cdot h \cdot g^{-1} \in g \cdot J \cdot g^{-1} = J$; similarly $g \cdot h \cdot g^{-1} \in K$, so $g \cdot h \cdot g^{-1} \in J \cap K$. A more conceptual answer to this question is that $J \cap K$ is the kernel of the homomorphism $g \to (g \cdot J, g \cdot K)$ of G into $(G/J) \times (G/K)$. Note that, more generally, the intersection of any nonempty collection of normal subgroups is a normal subgroup.

17. (a) Use the identity $(g \cdot h) \cdot H \cdot (g \cdot h)^{-1} = g \cdot (h \cdot H \cdot h^{-1}) \cdot g^{-1}$. Also if $H = g \cdot H \cdot g^{-1}$, then $g^{-1} \cdot H \cdot g = g^{-1} \cdot (g \cdot H \cdot g^{-1}) \cdot g = H$.
    (b) $\{ g \in G : g \cdot H \cdot g^{-1} = H \}$ is a subgroup [part (a)] of G containing A, and G is the smallest subgroup containing A. So $\{ g \in G : g \cdot H \cdot g^{-1} = H \} = G$, i.e., $g \cdot H \cdot g^{-1} = H$ for all $g \in G$ so that H is normal.

18. This exercise is just a restatement of Exercise 22 of § 12.5.

19. (a) $e \cdot K \cdot (1\ 3) \cdot K = e \cdot \{e, (1\ 2)\} \cdot (1\ 3) \cdot \{e, (1\ 2)\} = \{(1\ 3), (1\ 3\ 2), (1\ 2\ 3), (2\ 3)\}$, but cosets of K only have 2 members.
    (b) K is not normal. If it were, Theorem 2 would contradict part (a). Or check, for example, that $(1\ 3) \cdot K \neq K \cdot (1\ 3)$.

Section 12.7

1. (a) Yes.   (b) Yes, 1.   (c) No. Only 1 itself has an inverse.
    (d) It is a commutative monoid, but not a group.

2. (a) Yes.
    (b) Yes, the empty set $\emptyset$, since $A \oplus \emptyset = A$ for all $A \in \mathscr{P}(U)$.
    (c) Yes, $A \oplus A = \emptyset$. Elements of $\mathscr{P}(U)$ are their own inverses with respect to this operation.
    (d) $(\mathscr{P}(U), \oplus)$ is a group.

3. (a) Yes.   (b) Yes, the zero function.
   (c) Yes; compare Example 7(b).   (d) FUN($\mathbb{R}$,$\mathbb{R}$) is a group.

4. (a) Yes.
   (b) Yes. It is the constant function 1 with $1(x) = 1$ for all $x \in \mathbb{R}$.
   (c) No. A function that has the value 0 at some x has no inverse.
   (d) (FUN($\mathbb{R}$,$\mathbb{R}$),$\cdot$) is a monoid but not a group.

5. (a) No.   (b) Yes. It is the identity matrix $\begin{bmatrix} 1 & 0 \\ 0 & 1 \end{bmatrix}$.

   (c) No, the zero matrix has no inverse, for example.
   (d) ($\mathfrak{M}_{2,2}$,$\cdot$) is a monoid but not a group.

6. (a) badcab, cabbad, cababcdcabbad, abcdcababcd.
   (b) badbad, cabcabcab, λ.

7. (a) break, fast, fastfood, lunchbreak, foodfood.
   (b) breakfast $\neq$ fastbreak.
   (c) fastfast, foodfood, fastfastfoodbreakbreak, λ.

8. For example, $1/(1/2) = 2 \neq 1/2 = (1/1)/2$, so associativity fails.

9. (b) ($\mathbb{N}$,max) is a monoid because 0 is an identity: $\max\{n, 0\} = \max\{0, n\} = n$ for all
   $n \in \mathbb{N}$. ($\mathbb{N}$,min) has no identity and so it is not a monoid. To see this, assume that a
   number I in $\mathbb{N}$ satisfies $\min\{n, I\} = n$ for all $n \in \mathbb{N}$. In particular, we'd have
   $\min\{I + 1, I\} = I + 1$, i.e., $I = I + 1$, which is clearly false, no matter what I is.

10. (a) Since gcd(gcd(m,n),p) divides gcd(m,n), and since gcd(m,n) divides both m and
   n, gcd(gcd(m,n),p) divides m, n and p. Hence gcd(gcd(m,n),p) divides gcd(n,p)
   [the *greatest* common divisor of n and p], so it also divides gcd(m,gcd(n,p)).
   Similarly, gcd(m,gcd(n,p)) divides gcd(m,gcd(n,p)), so these two positive integers are
   equal. An alternative proof uses prime factorizations of integers.
   (b) Similar to part (a). Or use the observation that $\text{lcm}(m,n) = \dfrac{m \cdot n}{\gcd(m,n)}$ and part (a).
   (c) The semigroup in part (b) is a monoid, with identity 1: $\text{lcm}(n,1) = \text{lcm}(1,n) = n$ for
   all $n \in \mathbb{P}$. The one in part (a) has no identity element. If some integer I satisfied
   $\gcd(n,I) = n$ for all $n \in I$, then in particular we'd have $\gcd(2I,I) = 2I$, contrary to
   $\gcd(2I,I) = I$.

11. (a) $\mathbb{P}$.   (b) $\{0\}$.   (c) $\mathbb{Z}$.   (d) $\mathbb{P}$.   (e) $\mathbb{Z}$.
   (f) $\{2, 3, 4, 5, \ldots\} = \{n \in \mathbb{Z} : n \geq 2\}$.   (g) $18\mathbb{P} = \{18k : k \in \mathbb{P}\}$.

12. (a) $\{1\}$.   (b) $\{0\}$.   (c) $\{\pm1, \pm2, \pm4, \ldots\}$.   (d) $\mathbb{P}$.   (e) $\mathbb{Z}$.
   (f) $\{2^m 3^n : m \geq 0, n \geq 0$ and $m + n \geq 1\}$.

13. $\mathbb{P} = \{1\}^+$, $\{0\} = \{0\}^+$, $18\mathbb{P} = \{18\}^+$.

14. $\{1\} = \{1\}^+$, $\{0\} = \{0\}^+$. The other subsemigroups of $(\mathbb{Z}, \cdot)$ in Exercise 12 are not cyclic.

15. (a) $2\mathbb{N} = \{2k : k \in \mathbb{N}\}$.   (b) $\mathbb{Z}$.   (c) $\{0\}$.   (d) $\{1\}$.   (e) Σ^*.

16. (a) As in Exercise 11, $(\mathbb{P}, +)$ is a cyclic semigroup but its subsemigroup $\{2, 3\}^+ = \{n \in \mathbb{P} : n \geq 2\}$ is not.
   (b) $(\mathbb{Z}, +)$.
   (c) Any finite cyclic group $(\mathbb{Z}(p), +)$ will do.

17. (a) $\begin{bmatrix} 0 & 1 & 0 \\ 1 & 0 & 0 \\ 0 & 0 & 1 \end{bmatrix}$ and $\begin{bmatrix} 1 & 0 & 0 \\ 0 & 1 & 0 \\ 0 & 0 & 1 \end{bmatrix}$.

   (b) It consists of the six "permutation matrices," namely
   $$\begin{bmatrix} 1 & 0 & 0 \\ 0 & 1 & 0 \\ 0 & 0 & 1 \end{bmatrix}, \begin{bmatrix} 1 & 0 & 0 \\ 0 & 0 & 1 \\ 0 & 1 & 0 \end{bmatrix}, \begin{bmatrix} 0 & 1 & 0 \\ 1 & 0 & 0 \\ 0 & 0 & 1 \end{bmatrix}, \begin{bmatrix} 0 & 1 & 0 \\ 0 & 0 & 1 \\ 1 & 0 & 0 \end{bmatrix}, \begin{bmatrix} 0 & 0 & 1 \\ 1 & 0 & 0 \\ 0 & 1 & 0 \end{bmatrix}, \begin{bmatrix} 0 & 0 & 1 \\ 0 & 1 & 0 \\ 1 & 0 & 0 \end{bmatrix}.$$

   This semigroup is isomorphic to the group S_3 of all permutations of a 3-element set.

   (c) $\begin{bmatrix} 0 & 2 & 3 \\ 0 & 0 & 4 \\ 0 & 0 & 0 \end{bmatrix}, \begin{bmatrix} 0 & 0 & 8 \\ 0 & 0 & 0 \\ 0 & 0 & 0 \end{bmatrix}$ and $\begin{bmatrix} 0 & 0 & 0 \\ 0 & 0 & 0 \\ 0 & 0 & 0 \end{bmatrix}$.

18. (a) Subsemigroups (a) and (b) are groups. (c) is not.
   (b) Subsemigroups (a) and (c) are commutative.
   (c) By their definition, subsemigroups (a) and (c) are cyclic. Subsemigroup (b) isn't.

19. (a) $6\mathbb{P} = \{6k : k \in \mathbb{P}\}$.   (b) Yes, $6\mathbb{P} = \langle 6 \rangle^+$.
   (c) No. For example, 6 and 12 are not both powers of the same member of $6\mathbb{P}$, so cannot both lie in the same cyclic subgroup.

20. (a) The empty set $\emptyset$.   (b) No, it's not a subsemigroup.

21. (a) 60ℙ.

    (b) No. 60 and 120 are not both powers of the same member of 60ℙ.

    (c) 60 generates the *additive* semigroup 60ℙ.

22. Only (a) is a homomorphism from $(\mathbb{P},+)$ into $(\mathbb{P},\cdot)$. In (b), note $2 = \varphi(1+1) \neq \varphi(1)\cdot\varphi(1) = 1$. The image in (c) is not contained in $\mathbb{P}$. In (d) we have $4 = \varphi(1+1) \neq \varphi(1)\cdot\varphi(1) = 1$. In (e), $8 = \varphi(1+1) \neq \varphi(1)\cdot\varphi(1) = 16$.

23. Only the function φ in (a) is a homomorphism. Though it is one-to-one, it does not map $\mathbb{P}$ *onto* $\mathbb{P}$ and so it is not an isomorphism.

24. For $v,w \in \Sigma^*$, $\varphi(vw) = \text{length}(vw) = \text{length}(v) + \text{length}(w) = \varphi(v) + \varphi(w)$.

25. More generally, if $A \subseteq S$ then
$$\varphi(A^+) = \varphi(\{s : s \text{ is a product } a_1 \cdots a_n \text{ of members of } A\})$$
$$= \{\varphi(s) : s = a_1 \cdots a_n \text{ and } a_1, \ldots, a_n \in A\}$$
$$= \{t : t = \varphi(a_1) \cdots \varphi(a_n) \text{ and } a_1, \ldots, a_n \in A\} = \varphi(A)^+.$$

26. (a) More precisely, $\psi \circ \varphi$ is a homomorphism of $(S,\bullet)$ to $(U,\vartriangle)$.
$(\psi\circ\varphi)(s\bullet s') = \psi(\varphi(s\bullet s')) = \psi(\varphi(s)\,\Box\,\varphi(s')) = \psi(\varphi(s)) \vartriangle \psi(\varphi(s')) = (\psi\circ\varphi)(s) \vartriangle (\psi\circ\varphi)(s')$.

    (b) For $t,t' \in T$, $t\,\Box\,t' = \varphi(\varphi^{-1}(t))\,\Box\,\varphi(\varphi^{-1}(t')) = \varphi(\varphi^{-1}(t)\bullet\varphi^{-1}(t'))$, so $\varphi^{-1}(t\,\Box\,t') = \varphi^{-1}(t)\bullet\varphi^{-1}(t')$.

    (c) $S \simeq S$ since the identity map 1_S is an isomorphism. $S \simeq T$ implies $T \simeq S$ by part (b). And $S \simeq T \simeq U$ implies $S \simeq U$ by part (a).

27. (a) If z' is also a zero, then $z' = z \bullet z' = z$.

    (b) $(\mathbb{Z},\cdot)$.

    (c) Each $(\{0, 1, \ldots, n\}, \wedge)$ is an example, where $k \wedge \ell = \min(k,\ell)$. Another example is $(\mathscr{P}(U), \cap)$, with U a finite nonempty set; the zero element is $\varnothing$.

28. (a) Check that $\varphi(z)\,\Box\,\varphi(s) = \varphi(z) = \varphi(s)\,\Box\,\varphi(z)$ for every $s \in S$.

    (b) No. Consider $(S,\bullet) = (T,\Box) = (\mathbb{Z},\cdot)$ and $\varphi(n) = 1$ for all $n \in \mathbb{Z}$.

29. (a) $\varphi(S)$ is closed under products because $\varphi(s)\,\Box\,\varphi(s') = \varphi(s\bullet s') \in \varphi(S)$. And $\varphi(e)$ is an identity for $\varphi(S)$ because $\varphi(s)\,\Box\,\varphi(e) = \varphi(s\bullet e) = \varphi(s) = \varphi(e\bullet s) = \varphi(e)\,\Box\,\varphi(s)$ for all $\varphi(s)$ in $\varphi(S)$.

    (b) No. Consider, for example, any monoid $(S,\bullet)$ and define $\varphi: (S,\bullet) \to (\mathbb{Z},\cdot)$ by $\varphi(s) = 0$ for all $s \in S$.

Section 12.8

1. (a), (b), (d), (f).

   We supply explanations, thereby answering Exercise 2.

   (a) If $2m, 2n$ are in $2\mathbb{Z}$ [$m, n \in \mathbb{Z}$], so is their sum $2(m + n)$ and product $2(2mn)$.

   (b) $2\mathbb{R} = \mathbb{R}$ is closed under addition and multiplication.

   (c) $\mathbb{N}$ is not a subring, though it is closed under addition and multiplication. It isn't even an additive subgroup since $1 \in \mathbb{N}$ but $-1 \notin \mathbb{N}$.

   (d) Given $m + n\sqrt{2}$ and $p + q\sqrt{2}$ in the set, their sum $(m + p) + (n + q)\sqrt{2}$ and product $(mp + 2nq) + (np + mq)\sqrt{2}$ are also in the set.

   (e) The set is not closed under multiplication. It contains $1/2$ but not $(1/2) \cdot (1/2)$.

   (f) Consider $m/2^a$ and $n/2^b$ and assume $a \leq b$ so that $m/2^a = m'/2^b$ for $m' = m \cdot 2^{b-a}$. The sum has the form $(m' + n)/2^b$ and the product has the form $(mn)/2^{a+b}$. So the set is closed under sums and products.

2. See the answers to Exercise 1.

3. (a) $\varphi(f + g) = (f + g)(0) = f(0) + g(0) = \varphi(f) + \varphi(g)$ and $\varphi(f \cdot g) = (f \cdot g)(0) = f(0) \cdot g(0) = \varphi(f) \cdot \varphi(g)$.

   (b) $\varphi(1 + 1) = 4 \neq 2 = \varphi(1) + \varphi(1)$, so φ is not an additive homomorphism.

   (c) For $r, s \in \mathbb{R}$, $(\varphi(r + s))(x) = r + s = (\varphi(r))(x) + (\varphi(s))(x)$ for all $x \in \mathbb{R}$. Hence $\varphi(r + s) = \varphi(r) + \varphi(s)$. Likewise $\varphi(r \cdot s) = \varphi(r) \cdot \varphi(s)$.

   (d) φ is a homomorphism, as is easy to check.

   (e) $\varphi((n + 3\mathbb{Z}) \cdot (m + 3\mathbb{Z}))$

   $\qquad = \varphi(nm + 3\mathbb{Z})$ [definition of product in $\mathbb{Z}/3\mathbb{Z}$]

   $\qquad = 2nm + 6\mathbb{Z}$ [definition of φ]

   but

   $\varphi(n + 3\mathbb{Z}) \cdot \varphi(m + 3\mathbb{Z})$

   $\qquad = (2n + 6\mathbb{Z}) \cdot (2m + 6\mathbb{Z})$ [definition of φ]

   $\qquad = 4nm + 6\mathbb{Z}$ [definition of product in $\mathbb{Z}/6\mathbb{Z}$].

   Consequently, φ is not a ring homomorphism.

4. (a) $p(10) = 83017$, $q(10) = 10111$.

   (b) $r(x) = 8x^8 + 3x^7 + 8x^6 + 12x^5 + 18x^4 + 4x^3 + 8x^2 + 8x + 7$, and $r(10) = 839384887$.

5. (a) Evaluate both sides of the equation at a.

(b) $q_k(x) \cdot (x - a) = (x^{k-1} + ax^{k-2} + \cdots + a^{k-2}x + a^{k-1}) \cdot (x - a)$

$= x^k + ax^{k-1} + \cdots + a^{k-2}x^2 + a^{k-1}x - ax^{k-1} - \cdots - a^{k-2}x^2 - a^{k-1}x - a^k$

$= x^k - a^k$, or write it out with Σ-notation.

(c) By (b), $p(x) = \sum_{k=0}^{n} c_k x^k = \sum_{k=0}^{n} c_k \{q_k(x) \cdot (x - a) + a^k\} = \left(\sum_{k=0}^{n} c_k q_k(x) \right) \cdot (x - a) + \sum_{k=0}^{n} c_k a^k$.

Let $q(x) = \sum_{k=0}^{n} c_k q_k(x)$.

(d) $p(x)$ is in the kernel if and only if $p(a) = 0$. Use part (c).

6. $\{f \in \text{FUN}(\mathbb{R}, \mathbb{R}) : f(10) = 0\}$.

7. (a) $24\mathbb{Z}$.

(b) $6\mathbb{Z} + 8\mathbb{Z} = 2\mathbb{Z}$, since $2 = 6 \cdot (-1) + 8 \cdot 1 \in 6\mathbb{Z} + 8\mathbb{Z}$.

(c) $3\mathbb{Z} + 2\mathbb{Z} = 1\mathbb{Z} = \mathbb{Z}$, since $1 = 3 \cdot 1 + 2 \cdot (-1) \in 3\mathbb{Z} + 2\mathbb{Z}$.

(d) $6\mathbb{Z} + 10\mathbb{Z} + 15\mathbb{Z} = \mathbb{Z}$, since $6 + 10 - 15 = 1$.

8. We verify the first equality. In general, $x + y = 0$ implies $y = -x$ by the definition of additive inverse. Now $a \cdot b + (-a) \cdot b = [a + (-a)] \cdot b = 0 \cdot b$. Between Examples 2 and 3 it was shown that $0 \cdot b = 0$, and so $(-a) \cdot b = -a \cdot b$.

9. (a) Verify well-definedness directly, or apply Theorem 1 to the homomorphism $m \to (m \text{ MOD } 4, m \text{ MOD } 6)$ from $\mathbb{Z}$ to $\mathbb{Z}(4) \times \mathbb{Z}(6)$, as in Example 8. As noted in Exercise 12, it suffices to check that $m \to m \text{ MOD } 4$ and $m \to m \text{ MOD } 6$ are homomorphisms on $\mathbb{Z}(12)$. For example,
$$(m +_{12} n) \text{ MOD } 4 = m \text{ MOD } 4 +_4 n \text{ MOD } 4$$
because $m +_{12} n \equiv m + n \pmod{12}$ and hence also $m +_{12} n \equiv m + n \pmod 4$. Similarly $(m *_{12} n) \text{ MOD } 4 = m \text{ MOD } 4 *_4 n \text{ MOD } 4$.

(b) Kernel is $\{0\}$; φ is one-to-one.

(c) $(1,4)$ is one of the twelve; find another one.

(d) Just 9.

10. (a) Call the map ν. Then
$$\nu(r + s) = (r + s) + I = (r + I) + (s + I) = \nu(r) + \nu(s)$$
and
$$\nu(r \cdot s) = (r \cdot s) + I = (r + I) \cdot (s + I) = \nu(r) \cdot \nu(s),$$
using the definitions of addition and multiplication in R/I.

(b) $\nu(r) = I$ if and only if $r \in I$, so the kernel is I.

11. (a) Since $2 *_6 3 = 0$, 2 has no inverse.

   (b) Exhibit an inverse for each non-0 element. The inverses for non-0 elements in $\mathbb{Z}(5)$ can be read off of Figure 3 in § 3.6. [See Exercise 16 for the general argument.]

   (c) $F \times K$ isn't even an integral domain since $(1,0) \cdot (0,1) = (0,0)$.

12. $\varphi(r + s) = (\varphi_1(r + s), \varphi_2(r + s)) = (\varphi_1(r) + \varphi_1(s), \varphi_2(r) + \varphi_2(s)) = (\varphi_1(r), \varphi_2(r)) + (\varphi_1(s), \varphi_2(s)) = \varphi(r) + \varphi(s)$. Similarly, $\varphi(r \cdot s) = \varphi(r) \cdot \varphi(s)$.

13. (a) The kernel of φ is either F or $\{0\}$ by Example 6(b).

   (b) Since I is a subgroup of $(R,+)$, $\theta(I)$ is a subgroup of $(\theta(R),+)$. If $\theta(r) \in \theta(R)$ and $\theta(a) \in \theta(I)$, then $\theta(a) \cdot \theta(r) = \theta(a \cdot r) \in \theta(I)$ since $a \cdot r$ is in the ideal I. Similarly, $\theta(r) \cdot \theta(a) \in \theta(I)$, so $\theta(I)$ is closed under multiplication by elements of $\theta(R)$.

   (c) Use part (b) and Example 6(b).

14. (a) This is nearly obvious except for notation. Suppose R is an integral domain and $a \neq 0$. Then $a \cdot r = a \cdot s$ implies $r = s$, so $r \rightarrow a \cdot r$ is one-to-one.

   Suppose each map $r \rightarrow a \cdot r$ is one-to-one $[a \neq 0]$. Then $a \neq 0$ and $b \neq 0$ imply that $a \cdot b \neq a \cdot 0 = 0$, so R is an integral domain.

   (b) Suppose R is a field and $r \neq 0$. By part (a), $a \rightarrow r \cdot a$ is one-to-one. If this function didn't map R onto R, then the principal ideal $R \cdot r$ [remember, R is commutative] would be different from $\{0\}$ and R, contrary to Example 6(b).

   Suppose each $a \rightarrow r \cdot a$ maps R onto R, $r \neq 0$. Then given $r \neq 0$, $r \cdot r' = 1$ for some $r' \in R$. So R is a field.

   (c) Use (a) and (b). A one-to-one function of a finite set into itself is automatically an onto function, and hence a one-to-one correspondence.

15. (a) $I = 15\mathbb{Z}$.

   (b) See Example 8. Let $\varphi(m) = (m \text{ MOD } 3, m \text{ MOD } 4)$ for $m \in \mathbb{Z}(12)$. As in Exercise 9(a), φ is a homomorphism. Its kernel is easily checked to be $\{0\}$, so φ is an isomorphism of $\mathbb{Z}(12)$ onto $\mathbb{Z}(3) \times \mathbb{Z}(4)$.

16. (a) Apply the Euclidean algorithm of § 4.6 to the relatively prime integers k and p.

   (b) Apply MOD p to the equation in part (a) to get $k *_p (s \text{ MOD } p) +_p 0 *_p (t \text{ MOD } p) = 1$.

17. (a) $R \cdot 2 = \{a_0 + a_1 x + \cdots + a_n x^n \in R : \text{every } a_i \text{ is even}\}$,
   $R \cdot x = \{a_0 + a_1 x + \cdots + a_n x^n \in R : a_0 = 0\}$,
   $R \cdot 2 + R \cdot x = \{a_0 + a_1 x + \cdots + a_n x^n \in R : a_0 \text{ is even}\}$.

(b) Suppose $R \cdot p = R \cdot 2 + R \cdot x$ for some $p \in R$. Since $2 \in R \cdot p$, p must be constant, and since $x \in R \cdot p$, p is 1 or -1. But then $R \cdot p = R$, a contradiction.

18. As an additive group, $\mathbb{B} \times \mathbb{B}$ is simply $\mathbb{Z}(2) \times \mathbb{Z}(2)$. The non-0 elements form a group isomorphic to $\mathbb{Z}(3)$. It is generated by $(0,1)$ and also by $(1,1)$; $(1,0)$ is the multiplicative identity.

CHAPTER 13

This chapter contains topics that lead off in different directions from the main themes of Chapters 1 and 2. We have put the material at the end of the book because many teachers don't feel the need to cover this material and because it wasn't needed in earlier chapters. However, its level of sophistication matches the earlier chapters. Sections 13.1 and 13.2 can be covered after Chapter 2, while § 13.3 could be covered after Chapter 1, though it would be more natural after Chapter 5. Some students will enjoy studying § 13.3 on their own.

Section 13.1 introduces predicates as proposition-valued functions defined on sets called universes. The quantifiers $\forall$ and $\exists$ that were introduced informally in § 2.1 are now the key objects. One of the main messages here and in § 13.2 is that order matters for quantifiers. Exercise 15 is a tough one. We always assign it, but expect to help the students with it.

Section 13.2 illustrates the concept of tautology with the important example of DeMorgan's laws. Emphasize counterexamples as a way to disprove alleged tautologies, even though examples don't prove tautologies. Discuss how one can go about proving tautologies. Exercises 7-9 are good applications of DeMorgan's laws, as is Exercise 13. Exercises 6 and 10 give practice with counterexamples.

The purpose of § 13.3 is to expose students to the difference between countable and uncountable sets. There is not enough time or space to get very deeply into the subject, but students who have digested this section are ready to go on to more general questions about cardinal numbers. A good follow-up reference is *Set Theory–An Intuitive Approach*, by You-Feng Lin and Shwu-Yeng T. Lin, Houghton Mifflin Company, 1974. Exercise 9 is basic and should be assigned or gone over in class.

Section 13.1

1. (a) 0. Consider m odd.   (b) 0. Consider two different values of m.
   (c) 1.   (d) 0. Consider m odd or two different values of m.
   (e) 0. Consider $m = n = 0$.

2. (a) 0. Consider $x = 0$.   (b) 0. Consider $x = 0$ or any two different values of x.
   (c) 1.   (d) 0. Consider $x = y = 1$.   (e) 1. Consider $y = 0$.
   (f) 1. Consider $y = 0$.   (g) 1. $x = 8/5$ and $y = 6/5$.

(h) 0. By algebra $x^2 + y^2 + 1 = 2xy$ if and only if $(x - y)^2 = -1$.

3. (a) $\forall x \; \forall y \; \forall z \, [((x < y) \wedge (y < z)) \rightarrow (x < z)]$; universes $\mathbb{R}$.

(b) $\forall x \; \exists n \, [(n > x) \wedge (x > 1/n)]$; universe $(0, \infty)$ for x, universe $\mathbb{N}$ for n.

(c) $\forall m \; \forall n \; \exists p \, [(m < p) \wedge (p < n)]$; universes $\mathbb{N}$.

(d) $\exists u \; \forall n \, [un = n]$; universes $\mathbb{N}$.

(e) $\forall n \; \exists m \, [m < n]$; universes $\mathbb{N}$.

(f) $\forall n \; \exists m \, [(2^m \le n) \wedge (n < 2^{m+1})]$; universes $\mathbb{N}$.

4. (a) 1.   (b) 1.   (c) 0. Consider $m \ge n - 1$.

(d) 1. Consider $n = 1$.   (e) 0. Consider $n = 0$.

(f) 0. Consider $n = 0$.

5. (a) $\forall w_1 \; \forall w_2 \; \forall w_3 \, [(w_1 w_2 = w_1 w_3) \rightarrow (w_2 = w_3)]$.

(b) $\forall w \, [(\text{length}(w) = 1) \rightarrow (w \in \Sigma)]$.

(c) $\forall w_1 \; \forall w_2 \, [w_1 w_2 = w_2 w_1]$.

6. (a) 1.   (b) 1.

(c) If $a, b \in \Sigma$ with $a \ne b$, then $ab \ne ba$. The truth value is 0 if Σ has at least 2 elements and is 1 otherwise. In the English language, panfry $\ne$ frypan.

7. (a) x, z are bound; y is free.   (b) x is bound; y and z are free.

(c) Same answers as part (a).

8. (a) Both x and y are free.

(b) and (c) In fact, all expressions with one or two quantifiers, such as $\exists x \; \forall y \, [x + y = y + x]$, are true.

9. (a) x, y are free; there are no bound variables.

(b) $\forall x \; \forall y \, [(x - y)^2 = x^2 - y^2]$ is false. Consider $x = 0 \ne y$, for instance.

(c) $\exists x \; \exists y \, [(x - y)^2 = x^2 - y^2]$ is true.

10. (a) No. The proposition $\exists n \, [m + n = 7]$ is false for $m > 7$ and universe $\mathbb{N}$.

(b) Yes.

11. (a) No. $\exists m \, [m + 1 = n]$ is false for $n = 0$.   (b) Yes.

12. (a) No, since $1/(x^2 + 1) \notin \mathbb{N}$ for $x > 0$.

(b) Yes. Note that if x is rational, so is $1/(x^2 + 1)$.

(c) Yes.

13. (a) $\exists! x \, \forall y \, [x + y = y]$.   (b) $\exists! x \, [x^2 = x]$.

(c) $\exists! A \, \forall B \, [A \subseteq B]$. Here A, B vary over the universe of discourse $\mathscr{P}(\mathbb{N})$. Note that $\forall B \, [A \subseteq B]$ is true if and only if $A = \emptyset$.

(d) "$f: A \to B$" $\to \forall a \, \exists! b \, [f(a) = b]$. Here a ranges over A and b ranges over B. Alternative form: "$f: A \to B$" $\to \forall a \in A \, \exists! b \in B \, [f(a) = b]$.

(e) "$f: A \to B$ is a one-to-one function" $\to \forall b \, \exists! a \, [f(a) = b]$. Here a ranges over A and b ranges over B. One way to make this clear is to write $\forall b \in B \, \exists! a \in A \, [f(a) = b]$.

14. (a) 1. Use $x = 0$.   (b) 0. Both 0 and 1 are solutions.

(c) 1. Consider the empty set.   (d) 1. This follows from the definition of a function.

(e) 0. f need not map A **onto** B.

15. (a) True.

(b) False. The notation $\{0, 2, 4, 6, \ldots\}$ is deficient but is clearly meant to describe the set of all even nonnegative integers.

(c) False; e.g., 3 is in the right-hand set.

(d) False. The set described contains all odd positive integers as well as integers in A.

(e) False; the right-hand set is empty.

(f) False. The set described is $\mathbb{N}$ since $[2m = n \to m < 6]$ is trivially true for $m < 6$.

(g) True.   (h) True.   (i) True.   (j) True.   (k) True.

(ℓ) Vacuously true.   (m) True.

16. (a) There is an m in $\mathbb{N}$ such that, for every n in $\mathbb{N}$, if n is even, then $m + n$ is prime. In other words, there is an m in $\mathbb{N}$ such that $m + n$ is prime whenever n is an even nonnegative integer.

(b) For every n in $\mathbb{N}$ there is an m in $\mathbb{N}$ such that if n is not even, then $m + n$ is even.

(c) $\exists m \, \exists n \, [p(m) \wedge p(n) \wedge e(m + n) \wedge \neg (m = n)]$. Here the assumption that the two integers are different is implicit.

(d) $\forall m \, \forall n \, [(p(m) \wedge p(n) \wedge e(m + n)) \to (\neg (m = 2) \wedge \neg (n = 2))]$.

(e) $\forall m \, \forall n \, [(p(m) \wedge p(n)) \to \neg e(m + n)]$.

17. (a) 0. One of $m, m + 2$ and $m + 4$ is always a multiple of 3.

(b) 1. Consider any odd m.   (c) 1.   (d) 1.   (e) 0. Consider $3 + 5$.

Section 13.2

1. (a) Every club member has been a passenger on every airline if and only if every airline has had every club member as a passenger.

    (b) Some club member has been a passenger on some airline if and only if some airline has had some club member as a passenger.

    (c) If there is a club member who has been a passenger on every airline, then every airline has had a club member as a passenger.

2. (a) $\neg \forall x\, p(x)$.   (b) $\forall x\, \neg p(x)$.

    (c) As shown in Example 4(b), $\forall x\, \neg p(x)$ always implies $\neg \forall x\, p(x)$ for a nonempty universe. Thus (b) $\Rightarrow$ (a). (a) doesn't imply (b) since some, but not all, university professors like thrash metal.

    (d) There is no university professor who likes thrash metal if and only if every university professor does not like thrash metal.

    (e) Some university professor likes thrash metal if and only if not all university professors dislike thrash metal.

3. Rule 37b says that "There does not exist a yellow car" is logically equivalent to "Every car is not yellow." In fact, both are false. Rule 37c says that "Every car is yellow" is logically equivalent to "There does not exist a non-yellow car." Both are false. Rule 37d says that "There exists a yellow car" is logically equivalent to "Not every car is non-yellow." Both are true.

4. (a) 0. $\forall y\, [x \neq y]$ is false for every x but $\forall y\, \exists x\, [x \neq y]$ is true.

    (b) 1. This is the contrapositive of the predicate in Example 3(a).

    (c) 0 for each x; $p(x,y)$ is false for $y = x$.

    (d) 1.

5. (a) Rule 37d becomes $p(a) \vee p(b) \Leftrightarrow \neg(\neg p(a) \wedge \neg p(b))$. This is rule 8c of Table 1, § 2.2.

    (b) Rule 37b becomes $\neg(p(a) \vee p(b)) \Leftrightarrow (\neg p(a)) \wedge (\neg p(b))$, which is Rule 8a of Table 1 of § 2.2.

6. (a) Arrange for $p(x)$ and $q(x)$ to be true for disjoint sets of x's. For example, let $p(x)$ be "x is even" and $q(x)$ be "x is odd" with universe $\mathbb{N}$.

(b) Since $\neg (\forall x \,\exists y \, p(x,y)) \Leftrightarrow \exists x \,\forall y \,\neg p(x,y)$, the predicate $p(x,y)$ should be such that if U is nonempty there is some x for which $p(x,y)$ is true for all y, and some x for which $p(x,y)$ is false for all y. For example, let $p(x,y)$ be "$x + y = y$" with universes $\mathbb{N}$.

7. $\exists n\, [\neg \,(p(n) \to p(n + 1))]$ or $\exists n\, [p(n) \wedge \neg p(n + 1)]$.

8. $\forall x \,\exists y \,\forall z\, [(z > y) \wedge (z \geq x^2)]$.

9. (a) $\exists x \,\exists y\, [(x < y) \wedge \forall z\{(z \leq x) \vee (y \leq z)\}]$.
   (b) 1. For instance, let $z = (x + y)/2$.
   (c) 0; for example, $[x < y \to \exists z \,\{x < z < y\}]$ is false for $x = 3$ and $y = 4$.

10. (a) Any integer of the form $4n + 2$, such as 2 or 6.
    (b) Any example with $S \cap T \neq \emptyset$.
    (c) Let $k = 6$, for example.
    (d) A graph with 1 edge will suffice:
    (e) Actually, this statement is true.

11. One can let $q(x,y)$ be the predicate "$x = y$." Another way to handle $\exists x \, p(x,x)$ is to let $r(x)$ be the 1-place predicate $p(x,x)$. Then $\exists x \, r(x)$ is a compound predicate.

12. If U has one member, then $\forall x \, p(x)$ and $\exists x \, p(x)$ are either both true or both false.

13. $\exists N \,\forall n \,[p(n) \to (n < N)]$.

14. (a) Let the universe of discourse for the variable x be the set of negative integers, the universe of discourse for z be the set of positive integers, and let $\mathbb{Z}$ be the universe of discourse for y.
    (b) Let $\mathbb{Z}$ or $\mathbb{R}$ be the universe of discourse for all three variables x, y and z.

Section 13.3

1. (a) True   (b) False.   (c) False.   (d) False.
   (e) True. Compare Exercise 3 of § 1.3.

2. (a) True.   (b) True.   (c) False.

(d) False. It's countable, but it might not be infinite. It might even be empty.

(e) True.

3. (a) A function of the form $f(x) = ax + b$ will work if you choose a and b so that $f(0) = -1$ and $f(1) = 1$. Sketch your answer to see that it works. [For example, $f(x) = 2x - 1$ works.]

(b) Use g where $g(x) = 1 - x$.

(c) Modify suggestion for part (a). For example, $f(x) = 13x - 5$ works.

(d) Use $x \to 1/x$.

(e) Map $(1,\infty)$ onto $(0,\infty)$ using $h(x) = x - 1$ and compose with your answer from part (d) to obtain $h(1/x) = (1/x) - 1$.

(f) $f(x) = 2^x$, say. Sketch f to see that it works.

4. $f(n) = 2n - 2$ and $g(n) = 2n - 1$ work.

5. (a) Use the data

x	.1	.2	.3	.4	.5	.6	.7	.8	.9
$f(x)$	-8.89	-3.75	-1.90	-.83	0	.83	1.90	3.75	8.89

(b) The derivative is $(2x^2 - 2x + 1)/x^2(1 - x)^2$, which is positive on $(0,1)$ since $2x^2 - 2x + 1 = 2(x - \frac{1}{2})^2 + \frac{1}{2}$.

6. (a) - (f) are countable. (c), (e) and (f) are countably infinite. (g) is uncountable.

7. Only the sets in (b) and (c) are countably infinite.

8. If $m \le n$ there are $\dfrac{n!}{(n - m)!}$ one-to-one functions. If $m > n$ there are none.

9. (a) We may assume that S is infinite. Let $f: S \to T$ be a one-to-one correspondence where T is a countable set. There is a one-to-one correspondence $g: T \to \mathbb{P}$ since T is countable. Then $g \circ f$ is a one-to-one correspondence of S onto $\mathbb{P}$.

(b) Suppose f is a one-to-one correspondence of S onto the uncountable set T. If S were countable, then $f^{\leftarrow}$ would be a one-to-one correspondence of T onto the countable set S, so that T would be countable by part (a), a contradiction.

10. By construction, $f(n + 1) \neq f(k)$ for $1 \leq k < n + 1$, so no two values of f are equal, and f is one-to-one.

   To show that f maps $\mathbb{P}$ onto A, consider $m \in A$. Some member of the infinite set $\{f(1), f(2), \ldots\}$ of positive integers is larger than m; say $m < f(n)$ for some $n \in \mathbb{P}$. Then m is in $\{f(1), \ldots, f(n-1)\}$, since otherwise $f(n)$ would not be the smallest member of $A \setminus \{f(1), \ldots, f(n-1)\}$. Thus $m = f(k)$ for some $k \in \mathbb{P}$. That is, every member of A is in the image of f.

11. (a) Apply part (b) of the theorem to $S \times T = \bigcup_{t \in T} (S \times \{t\})$. Each $S \times \{t\}$ is countable, since it is in one-to-one correspondence with the countable set S.

   (b) For each $t \in T$, let $g(t)$ be an element in S such that $f(g(t)) = t$. Show that g is one-to-one and apply part (a) of the theorem.

   (c) By part (a), $\mathbb{Z} \times \mathbb{P}$ is countable. Since f maps $\mathbb{Z} \times \mathbb{P}$ onto $\mathbb{Q}$, $\mathbb{Q}$ is countable by part (b).

12. If S and T have the same size, there is a one-to-one function f from S onto T. Use f to define a function $f^*: \mathcal{P}(S) \longrightarrow \mathcal{P}(T)$ by $f^*(A) = f(A) = \{f(a) : a \in A\}$ for each subset A of S. The inverse of f defines the inverse of f^* similarly, so f^* is a one-to-one function mapping $\mathcal{P}(S)$ onto $\mathcal{P}(T)$.

13. (a) For each f in $\mathrm{FUN}(\mathbb{P}, \{0, 1\})$, define $\phi(f)$ to be the set $\{n \in \mathbb{P} : f(n) = 1\}$. If $f, g \in \mathrm{FUN}(\mathbb{P}, \{0, 1\})$ and $f \neq g$, then there exists $k \in \mathbb{P}$ so that $f(k) \neq g(k)$. Then k belongs to $\{n \in \mathbb{P} : f(n) = 1\}$ or $\{n \in \mathbb{P} : g(n) = 1\}$ but not both. Hence $\phi(f) \neq \phi(g)$; this shows that ϕ is one-to-one. ϕ maps onto $\mathcal{P}(\mathbb{P})$ because given $A \in \mathcal{P}(\mathbb{P})$ its characteristic function χ_A belongs to $\mathrm{FUN}(\mathbb{P}, \{0, 1\})$ and $\phi(\chi_A) = A$.

   (b) Use Example 2(a) and Exercise 9.

14. Choose a member from each subset in the family. The set of chosen elements is in one-to-one correspondence with the family and is countable by part (a) of the theorem. Apply Exercise 9(a).

15. For the inductive step, use the identity $S^n = S^{n-1} \times S$.

16. (a) $8 = 2^3$, so $f(\frac{1}{8}) = 2^5 = 32$. $9 = 3^2$, so $f(\frac{1}{9}) = 3^3 = 27$. $10 = 2 \cdot 5$, so $f(\frac{1}{10}) = 2 \cdot 5 = 10$. $100 = 2^2 \cdot 5^2$, so $f(\frac{1}{100}) = 2^3 \cdot 5^3 = 1000$. $f(\frac{21}{20}) = 3^2 \cdot 7^2 \cdot 2^3 \cdot 5 = 17{,}640$.

   (b) $f(\frac{1}{23}) = 23$. $f(\frac{1}{12}) = 24$. $f(5) = 25$. $f(\frac{1}{26}) = 26$. $f(\frac{1}{9}) = 27$. $f(\frac{2}{7}) = 28$.

(c) Consider a positive integer N and write it as a product of primes: $N = p_1^{2m_1} \cdots p_k^{2m_k} \cdot q_1^{2n_1 \cdot 1} \cdots q_\ell^{2n_\ell \cdot 1}$ where we have written the even powers of primes first. All p_i's are distinct, all q's are distinct, and no p_i equals any q_j. If $m = p_1^{m_1} \cdots p_k^{m_k}$ and $n = q_1^{n_1} \cdots q_\ell^{n_\ell}$, then $f(\frac{m}{n}) = N$. Moreover, this is the only fraction in $\mathbf{Q}^+$ that maps to N because the factorization of a positive integer into a product of primes is unique (except for order).